Looking at Ribozymes

Nucleic Acids Set

coordinated by
Marie-Christine Maurel

Volume 2

Looking at Ribozymes

Biology of Catalytic RNA

Benoît Masquida

with the participation of
Fabrice Leclerc

WILEY

First published 2023 in Great Britain and the United States by ISTE Ltd and John Wiley & Sons, Inc.

Apart from any fair dealing for the purposes of research or private study, or criticism or review, as permitted under the Copyright, Designs and Patents Act 1988, this publication may only be reproduced, stored or transmitted, in any form or by any means, with the prior permission in writing of the publishers, or in the case of reprographic reproduction in accordance with the terms and licenses issued by the CLA. Enquiries concerning reproduction outside these terms should be sent to the publishers at the undermentioned address:

ISTE Ltd	John Wiley & Sons, Inc.
27-37 St George's Road	111 River Street
London SW19 4EU	Hoboken, NJ 07030
UK	USA
www.iste.co.uk	www.wiley.com

© ISTE Ltd 2023

The rights of Benoît Masquida and Fabrice Leclerc to be identified as the authors of this work have been asserted by them in accordance with the Copyright, Designs and Patents Act 1988.

Any opinions, findings, and conclusions or recommendations expressed in this material are those of the author(s), contributor(s) or editor(s) and do not necessarily reflect the views of ISTE Group.

Library of Congress Control Number: 2023946400

British Library Cataloguing-in-Publication Data
A CIP record for this book is available from the British Library
ISBN 978-1-78630-977-8

Contents

Foreword vii

Preface xi

Chapter 1. Fundamentals of RNA and Ribozyme Structure 1

   1.1. Sequences and secondary structures 2
   1.2. RNA folding, tertiary structures and 3D 8
      1.2.1. Secondary structures and RNA folding 8
      1.2.2. The pseudoknot 12
      1.2.3. The loop E 14
      1.2.4. The k-turn 16
      1.2.5. Tetra-loop receptors 18
      1.2.6. The A-minor pattern 20
      1.2.7. Comparative analysis of sequences 24

Chapter 2. Ribozymes and the "Central Dogma" of Molecular Biology 29

   2.1. The discovery of RNA catalysis and the central dogma of
   molecular biology. 29
   2.2. In search of the primordial polymerase 34

Chapter 3. The Discovery of Ribozymes 37

   3.1. The discovery of catalysis by autocatalytic introns 38
   3.2. The discovery of RNA catalysis of RNase P 42
   3.3. The first consequences of these discoveries. 46
      3.3.1. The ribosome, a long-ignored ribozyme 47
      3.3.2. The modified bases. 52

3.4. The spliceosome, another ribozyme 55
3.4.1. Nucleolytic ribozymes 56

Chapter 4. Ribozyme Engineering and the RNA World 73

4.1. Classification of ribozymes 75
4.2. Classification of ribozymes according to catalytic mechanism 76

Chapter 5. Structures of Ribozymes 81

5.1. Structures and catalytic mechanisms of ribozymes 81
5.1.1. Hammerhead ribozymes 81
5.1.2. The example of the hairpin ribozyme 87
5.1.3. The example of the glmS ribozyme 89
5.2. An example of catalysis control: lariat-capping ribozyme 91

Chapter 6. Evolution of the Vision of the Catalytic Mechanisms of Ribozymes, the Hammerhead Ribozyme 99

6.1. Chemistry and catalysis: between general acid/base and metal cations 99
6.2. Difficulties in interpreting catalysis data 103

Chapter 7. The Distribution of Ribozymes in Living Organisms and Molecular Adaptations during Evolution 111

7.1. Ubiquitous ribozymes 111
7.2. Selection pressures at work in ribozyme shaping 113
7.3. Ribozymes in cellular processes: from viroids to eukaryotes 117
7.4. Very human ribozymes 123

Conclusion 133

References 135

Index 165

Foreword

Ribozymes are enzymes composed of RNA instead of proteins. Proteins have 20 different amino acid side chains, covering a wide range of aqueous chemistry. In contrast, RNA has fewer resources, with only four nucleic bases with a 2' hydroxyl group. However, its polyanionic nature allows it to recruit polycations, and thus bind hydrated metal ions that can also participate in chemical catalysis. Despite their limited catalytic resources, ribozymes are capable of accelerating a chemical reaction by a million times or more.

Until the 1980s, the only known biocatalysts were proteins. This all changed when Tom Cech discovered that splicing of the *Tetrahymena* group I intron was autocatalyzed. At the same time, Norm Pace and Sidney Altman were studying RNase P, which processes the 5' ends of RNA transfer, and they discovered that the active component was actually an RNA molecule, not a protein subunit. These discoveries opened up a whole new field of biological chemistry, challenging the chemist to understand how RNA could act as an enzyme with so few chemical resources.

Why are we interested in ribozymes? And why are they important in biology? First, they were most likely the only biological catalysts at an early stage in the development of life on Earth. There is a big "chicken and egg" problem in starting life, with both proteins and nucleic acids being needed. Evolution could not begin until there was genetic coding, replication and formation of biological catalysts. However, the RNA world hypothesis assumes that early life would have begun with RNA as both a genetic and catalytic reserve. This would therefore require ribozymes to catalyze primitive metabolism at this stage.

Proteins have a number of advantages over RNA as catalysts. It is therefore likely that the world of RNA would have been overtaken relatively quickly by a world of proteins. Nevertheless, ribozymes still exist in contemporary biology. This is the second reason why ribozymes are important. RNA catalyzes what is perhaps the most important reaction in the cell, the condensation of amino acids to form polypeptides in the large ribosomal subunit. That is, the ribosome is a ribozyme! The splicing of mRNA by the spliceosome is also catalyzed by RNA from the spliceosome. In addition, RNase P is involved in the maturation of tRNAs in all areas of life.

A group of nucleolytic ribozymes are the site-specific nucleases. These are generally part of the smaller catalytic RNA group. They were discovered in Australia, as they are present in viruses significant in the agricultural industry. In Adelaide, I met Bob Symons, who discovered the hammerhead ribozyme relatively late in life, and I thought he was more interested in grape growing than molecular biology! There are nine ribozymes in this class, the most recent of which were found through the application of directed bioinformatics in Ron Breaker's lab. Some of the nucleolytic ribozymes have been found to be widespread, including in the human genome. In most cases, we do not yet know their function, but they are probably important.

This book presents a synthesis of the current state of knowledge in the field of ribozymes. So what are the big questions? Obviously, a major question is how can such major feats of chemical catalysis be achieved? This question can be asked at almost any level, and I am not sure we fully understand a protein enzyme. But in general terms, we have a general understanding, and for some ribozymes, our description of their catalytic mechanism is very good. Of course, any catalyst must ultimately stabilize the transition state of the reaction, but how do we achieve this? There are two major processes used by ribozymes, which divide them into two classes. Self-splicing introns and RNase P use hydrated metal ions to activate the nucleophile and organize the transition state. In contrast, nucleolytic ribozymes use general acid–base catalysis, with a major role for catalytic nucleobases.

A second major question is the extent to which RNA-based catalysis is used in current biology, and whether there are major classes of ribozymes that have escaped our discovery so far. Clearly, an RNA world would have required enzymes that catalyze a much wider range of chemical reactions, including "difficult" ones like C–C bond formation. It is very exciting to speculate on what solutions might exist, but how would we find them and where should we look? One way to find out what RNA might be capable of is to use in vitro selection methods. There has been a recent surge of interest in RNA species that can accelerate methyl transfer reactions, for example.

One way biology could have overcome the lack of chemical functionality of RNA might have been to use small molecules as cofactors, much like proteins use coenzymes. An example already exists with the nucleolytic ribozyme GlmS, which uses the captive ammonium group of glucosamine as the general acid. RNA is excellent at capturing small molecules, and here riboswitches can show the way. Many classes of riboswitches interact with potent coenzymes (suggesting an ancient origin tracing back to the RNA world), and we might imagine that such RNA molecules could relatively easily be converted into ribozymes using their coenzymes in catalysis.

Thus, ribozymes are important, and present a challenge to the chemist to understand their mechanisms, and at the same time perhaps shed a light on biocatalysis in general. In many cases, they also present interesting questions about their biology and the extent to which they have evolved in directions that we have not yet explored in depth. I hope that new generations will take up this challenge, and this book will provide an excellent starting point for that journey.

David M.J. LILLEY
University of Dundee

Preface

Ribozymes are enzymes composed of nucleic acids and in nature more specifically of RNA. They are at the origin of life as Thomas Cech and Sidney Altman proposed in an emblematic book entitled *The RNA World: The Nature of Modern RNA Suggests a Prebiotic RNA World* (Gesteland and Atkins 1993). The expansion of life indeed supposes the enzymatic copy of a pre-genetic material. Thus, nucleic acids combine two properties that proteins do not have. They pair via nucleobases by forming double helices in which one strand is a negative copy of the other, in reference to analog photography. In addition, the catalytic properties of their chemical groups, even if limited, accelerate millions of times over the transesterification reactions used for the synthesis of this copy. It is thus possible to design a self-replicating ribozyme that represents the archetype of an autonomous prebiotic system. The development of this primordial system led to the transmission and translation of the genetic material as we know it today, with the transcription of DNA into RNA followed by its translation into proteins.

In biology, there are numerous reminiscences of this system. The translation of messenger RNAs into proteins via the ribosome is probably the example that comes most immediately to mind since the ribosome is a ribozyme. Thus, the synthesis of the peptide bond of proteins, transpeptidylation, depends directly on RNA catalysis. The same is true for splicing by both autocatalytic introns, discovered by Thomas Cech, and spliced introns by the spliceosome. The vitamins and enzymatic cofactors of proteins are frequently nucleotide derivatives (FMN, FAD, NAD, vitamin

B12, coenzyme A, ATP, SAM, to constitute a non-exhaustive list)[1]. These nucleotide derivatives specialized hundreds of millions of years ago upon contact with RNAs to perform specific chemical reactions beyond their reach, thus diversifying the catalytic repertoire of ribozymes and also leading to proteins emerging in the chemistry of life. This dependence of protein enzymes on these cofactors, many of which are derived from nucleic acids, as well as the relatively easy possibility of "selecting" in vitro RNA sequences capable of recognizing these same cofactors suggest that in the "RNA world" the catalytic diversity was greater than that visible today. This also means that the proteins offer reactional diversity, admittedly greater than that of RNAs, but still insufficient to do without these cofactors. Why would proteins have taken precedence over RNA at some point? Probably for reasons of genome stability and catalytic speed, but these are only partial answers. More recently identified ribozymes also suggest that they act in the developmental programs of organisms, cell differentiation, memory or stress. It is safe to say that our view of ribozymes in biology is sufficiently fragmentary that we have not yet grasped the diversity of their roles.

Hence, it became necessary to publish a book presenting this field of biology, in order to draw up an inventory allowing the greatest number of people to understand the importance of this group of molecules. Our ambition is to give the reader, from the beginning of higher education in life sciences to the confirmed researcher wishing to acquire structural knowledge, the possibility to quickly familiarize with this flourishing field that seems to have gone a bit out of fashion. However, this is not the case! Modern biology methods, and in particular high-throughput sequencing, are giving a second wind to ribozyme research. Indeed, since their discovery, the emblematic ribozymes have been mainly studied at a mechanistic level because their role was well established. But the presence of ribozymes in genomes points to many unsuspected roles that need to be studied further in order to gain a deeper understanding of the diversity and complexity of life. The preponderance of transcribed but untranslated sequences in genomes indicates an essential role for RNAs among which still unknown ribozymes certainly remain to be identified and studied.

This book covers four topics. Chapter 1 provides structural knowledge on nucleic acids in order to understand the following parts. Chapters 2 and 3

---

1 FMN: flavin mononucleotide; FAD: flavin adenine dinucleotide; NAD: nicotinamide; ATP: adenosine triphosphate; SAM: S-adenosyl methionine.

describe the history of the discovery of ribozymes and review the state of knowledge about ribozymes, which have been studied since the 1980s. Catalysis by ribozymes is discussed in Chapters 4–6 and allows us to visualize the strategies used by these molecules to enable transesterification reactions. Finally, in Chapter 7, the roles of ribozymes are discussed in relation to the context in which they have been identified, thus whetting our curiosity when it is realized that more questions than answers are provided.

We hope you enjoy reading!

Acknowledgments

We thank Professor David M.J. Lilley (University of Dundee, United Kingdom) and Drs François Michel (Institut de systématique, évolution et biodiversité, Museum d'histoire naturelle, Paris) and Maria Costa (I2BC, Paris Saclay) for critically reviewing this manuscript. We would also like to thank Marie-Christine Maurel for giving us the opportunity to write this book and for her unfailing encouragement throughout the project.

More indirectly, we would also like to thank our mentors and researcher colleagues, as well as teacher-researchers and technicians who contribute to the advancement of knowledge. Research is a team effort where the notion of community is paramount. Community is understood here as a structure in which all members work in synergy with a common interest beyond particular or individual interests. Scientific research is a risk-taking activity in which even the instigators are unable to foresee the steps that will lead to discovery. This particular context can only foster discovery if researchers are free to focus their thoughts on their object, without these being polluted by materialistic considerations. "Research cannot be programmed, it takes time and cannot be done with an uncertain status," to quote Rose Katz, a researcher at INSERM who passed away in 2022.

The three-dimensional molecule figures present in most of the images in this book were created using PyMol software (Schrodinger 2010).

September 2023

1
Fundamentals of RNA and Ribozyme Structure

This chapter is intended to familiarize the reader with the structure of RNAs. Understanding the structural basis of RNAs is a prerequisite for the study of ribozymes and RNA-mediated catalysis. This section shows how nucleotide stereochemistry guides the structuring of helices and consequently the addition of functional motifs that give this polymer its folding and interaction properties, as well as its catalytic properties.

It is important to understand that all biological mechanisms rely on the interaction capabilities of structured molecules. The structure of biomolecules is therefore a fundamental aspect for the understanding of biology. The figures in this book are therefore often developed from experimental structures obtained by radio-crystallography or electron microscopy. Visualizing a biological mechanism through the molecular structures involved allows a better understanding of the actions of the different partners and the domains that compose them. It is then possible to deduce the mechanisms of chemical reactions and also to establish evolutionary relationships between homologous molecules of different organisms that perpetuate these mechanisms while adapting to different selection pressures resulting from distinct ecological constraints.

For a color version of all the figures in this chapter, see www.iste.co.uk/masquida/ribozymes.zip.

1.1. Sequences and secondary structures

RNA (ribonucleic acid) is one of the three main biological polymers with DNA (deoxyribonucleic acid) and proteins. RNA adopts complex structures thanks to the physicochemical properties of the four main nucleotides from which it is assembled. The nucleotides are composed of a ribose-phosphate part and an aromatic part, the nucleobase, attached to the ribose. The base is composed of one or two fused aromatic rings containing imines and ethylenic carbons decorated by exocyclic amines and/or carbonyl groups. The ribose is also substituted with a phosphate group (Figure 1.1(a)). The phosphate group gives each nucleotide a negative charge. The "backbone" of the polymer is thus a polyanion. The decorations of the aromatic bases generate an electrostatic profile specific to each one that allows for the local appearance of negative (δ^-) and/or positive (δ^+) partial charges. RNAs are therefore not simple polyanions. The electrostatic profile of the bases confers on nucleotides interaction properties between them, as well as with the ions and water molecules that solvate them (Auffinger et al. 2016; D'Ascenzo and Auffinger 2016; Leonarski et al. 2017, 2019). Nucleotides tend to stack and form right-handed double-stranded helices promoted by the establishment of hydrogen bonds between bases. The 5'–3' orientation of the strands is opposite. The strands are therefore antiparallel (Figure 1.1(b)).

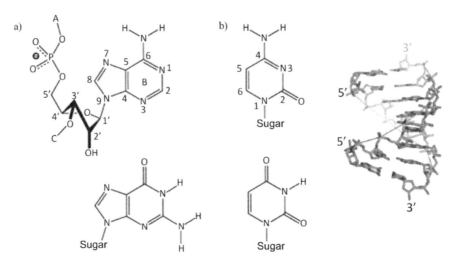

Figure 1.1. *The four nucleotides of RNA form antiparallel double-stranded helical structures*

COMMENTARY ON FIGURE 1.1.– *a) Two purines (N) on the left and two pyrimidines on the right (Y). The ribose and the negatively charged phosphate group are shown on the adenine along with the numbering of the atoms according to the IUPAC (International Union of Pure and Applied Chemistry) system. The ribose atoms are numbered from 1' to 5' and the base atoms from 1 to 9 for a purine (R) and from 1 to 6 for a pyrimidine (Y). The oxygen atom of the ribose ring corresponds to the hydroxyl group carried by the C4' and therefore has the same number (O4'). The situation is identical for the hydroxyl groups carried by C2' and C3'. Uracil and cytosine have an O4 or N4 group, respectively. The polynucleic acid parent chains are oriented from 5' to 3'. The present example gives the sequence 5'-ABC-3'. b) Since each phosphate group carries a negative charge, a polynucleotide is a polyanion whose structure mostly forms right-handed antiparallel double-stranded helices. The bases pair up to form plateaus of bases that stack with each other.*

Despite different electrostatic properties, the stereochemistry of each nucleotide is identical. The βD-ribofuranose isomer[1], hereafter simply referred to as ribose, adopts an envelope (E) or twist (T) fold depending on whether four or three ring atoms define a plane, respectively. This plane is oriented using the only non-asymmetric carbon atom outside the ribose ring, the C5'. When the ribose is drawn with the C5' above the plane and to the left, the C2' and C3' carbon atoms point forward and the O4' points backward (Figure 1.2(a)). This configuration is the only one found in nature and gives an idea of the intensity of selection pressures that led to the emergence of these stereoisomers in biological nucleic acid synthesis pathways. The base is branched at the C1' position and points to the same edge as the phosphate group. All nucleotides are therefore superimposable to each other by their ribose-phosphate part. The strands (chains) of nucleotides are naturally structured into helices that interact with each other to form double-stranded helices characterized by grooves of different morphologies (Figure 1.2(b)). In fact, the path between the riboses of a base pair is shorter on one edge of the helix than on the other, which gives rise to the notion of major and minor grooves. In RNA, the minor groove is wide and shallow, and the major groove is narrow and deep. Width and depth are not independent. A deep groove is narrow, and a wide groove is shallow. The

---

1 βD-ribofuranose is an enantiomer of ribose, a 5-carbon aldose that cyclizes by attack of the C4'-carried hydroxyl group on the C1' aldehyde.

proteins that interact with RNA therefore tend to interact with the shallow groove.

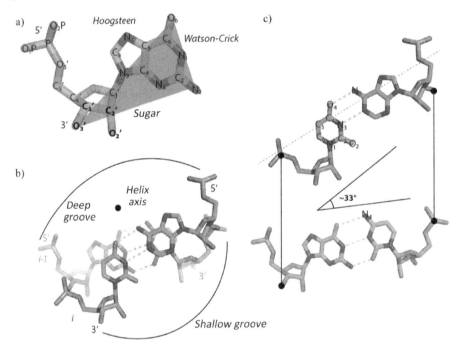

Figure 1.2. *Nucleotides and base pairs generating antiparallel double-stranded helices*

COMMENTARY ON FIGURE 1.2.– *a) Nucleotides adopt a precise conformation within the helices. The three edges of the nucleobases define the Watson–Crick, Hoogsteen and Sugar edges schematized by the edges of a right triangle. The atoms are numbered as shown. O1P and O2P are not linearly integrated into the 5'–3' ribose-phosphate backbone. The phosphate moiety is prochiral because the tetrahedral phosphorus has two identical moieties (O1P corresponds to the Pro-S stereoisomer and O2P corresponds to the Pro-R with reference to the R and S stereodescriptors). b) RNA helices are formed by the stacking of base pairs formed via Watson–Crick edges. Watson–Crick helices have an antiparallel orientation of the 5'–3' strands. If the 5' end of one strand is behind the plane of the sheet, then the 5' end of the other strand is above the plane. c) The permutations of these base pairs are isosteric, i.e. their riboses are superimposable 2 to 2. The orange dashed line collinear to the hydrogen bonds verifies whether the pairing is in the cis (both riboses are on the same edge) or trans configuration (both riboses are on opposite sides of this line). The O3'–P bonds between two adjacent riboses induce a right-handed rotation of about 33° between two consecutive base pairs. The A-form RNA helices therefore rotate to the right.*

Five important properties of RNA follow from this absolute configuration.

The planar structure of nucleobases defines three hydrogen bond acceptor and/or donor edges, Watson–Crick (W or WC), Hoogsteen (H) and sugar (S). The W edge is responsible for the base pairs identified in the DNA double helix model by James Watson and Francis Crick in 1953 (Watson and Crick 1953). Karst Hoogsteen (1963) was the first to observe pairings involving the N7 and N6 positions of adenines, thus giving his name to this edge of the nucleobases:

– The O2' group plays a prominent role in interactions with the sugar edge and, as we will see later, in catalysis. Since DNA is free of the O2' group, its structural repertoire is consequently not as rich as that of RNA, as is its chemical reactivity.

– The nucleotides are linked to each other from the 5' position to the 3' position. The ribose-phosphate backbone is thus polarized. In projection on the axis of the helix, the nucleotide *i-1* which precedes the nucleotide considered *i* is thus "above" the plane of the ribose.

– The plane of the base is perpendicular to the plane of the ribose and to the helix axis (Figure 1.2(b)).

– The grooves of the helix have different widths and depths. The deep groove is less accessible because it is narrow due to the stacking of the base plateaus. On the contrary, the shallow groove is very accessible. The S sides of the bases are therefore more accessible for another macromolecule than the H sides. The W sides are involved in pairing and are therefore less frequently involved in interactions with other molecular partners.

These five properties result in the formation of anti-parallel double-stranded helices. In a multi-nucleotide chain, the bases tend to stack with each other (stacking interactions), providing the edges capable of forming hydrogen bonds with the opportunity to interact with another RNA chain. For a helix to form, the interacting bases must be complementary. The Watson–Crick edges of a purine A or G always form a base pair with the

Watson–Crick edges of a pyrimidine U or C, respectively, by establishing hydrogen bonds. In RNAs, G can also interact with U by forming two hydrogen bonds. This particular conformation is called wobble because it induces a shift towards the deep groove of the pyrimidine. This geometry stabilizes helices whose strands are not totally complementary. This geometry was proposed long before the first crystallographic structures by Crick (1966). This particular geometry is of great importance in biology both structurally and in terms of the genetic code that converts nucleic acid sequences into proteins, as we will see later.

In Watson–Crick base pairs, the riboses are on the same edge *cis* to a line collinear to the hydrogen bond axis (Figure 1.2(c)). This configuration results in the 5'–3' polarities of the strands of a helix being opposite. The helix is said to be antiparallel, and the positions of the elements of the ribose-phosphate backbone are always equivalent whatever the pairings A–U, U–A, G=C, C=G put in place. These pairings are said to be isosteric, i.e. of the same volume. Each pair of bases is therefore superimposable on the previous and the next, which allows them to be stacked. In addition, the βD stereoisomer of the ribose induces a right-hand twist of the helix of about 33° per helix turn (Figure 1.2(c)). Thus, the common part of the nucleotides forms the ribose-phosphate backbone at the periphery of the helix and the distinctive part that is the nucleobase forms the core. The nucleotides thus interact between them by their distinctive part, the nucleobase. The isostery between *cisW* pairings also means that the helix structure is conserved regardless of the W pairings in place. The opposite occurs in proteins where the alpha helix and beta strand secondary structure domains assembled in a sheet establish hydrogen bonds between the backbone atoms containing the peptide bonds, while the distinctive side chains point to the periphery. Therefore, the nucleotide sequence gives immediate structural information that the protein sequence does not.

Another important concept is the dihedral angle. A dihedral is formed by the intersection of two half-planes (Figure 1.3). The orientation of one half-plane with respect to the other corresponds to the value of the dihedral angle. If the boundary between the two half-planes corresponds to a bond between two atoms, each atom admits at least one other bond belonging strictly to one of the half-planes (Figure 1.3). Four atoms (1, 2, 3, 4) define a dihedral angle between atoms 2 and 3. In nucleic acids, these dihedral angles are denoted by Greek letters. Only one dihedral angle is located outside the ribose phosphate chain, χ (pronounced "khi"). In general, χ is fixed at ~180°

(this conformation is also called *anti*). Cases where χ adopts the opposite value of ~0° (*syn*) are rare and related to a structural context that favors interactions with neighboring nucleotides. In the remainder of this work, the considerations developed assume that nucleotides adopt the *anti* conformation unless otherwise specified. In general, the dihedral angles adopt values that belong to domains favored by the steric hindrance of the groups, on the one hand, and the conformation, on the other hand. These domains are located in the +60° (*gauche$^+$*, g^+), -60° (*gauche$^-$*, g^-) and 180° (*trans*) regions. Thus, β, δ and ε adopt a single triplet of values, *trans, g^+, trans*, respectively, while γ preferably adopts the g^+ or g^- conformations. Only α and ζ adopt the three values g^-, g^+ or *trans*. Nucleotides can thus be considered as well-defined conformation blocks, whereas six torsion angles should produce an infinity of conformations. This is the Levinthal paradox (Levinthal 1968; Zwanzig et al. 1992). The variations taking place around the phosphate group result from the local conformation necessary for the folding of the RNA molecules and therefore of their final structure (Sundaralingam 1973).

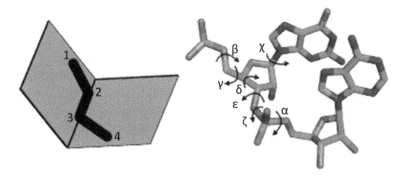

Figure 1.3. *Illustration of the notion of a dihedral angle and the different dihedrals present in an RNA dinucleotide*

COMMENTARY ON FIGURE 1.3.– *The atoms involved in determining the values (°) of the dihedral angles are the following: α (O3'–P–O5'–C5'), β (P–O5'–C5'–C4'), γ (O5'–C5'–C4'–C3'), δ (C5'–C4'–C3'–O3'), ε (C4'–C3'–O3'–P), ζ (C3'–O3'–P–O5'), χ (O4'–C1'–N9(R)/C1(Y)–C4(R)–C2(Y)). By looking at the angle in the axis of atoms 3 and 2, we determine which angular sector to turn in order to eclipse atom 1 with atom 4. Turning clockwise gives the positive (+) direction.*

1.2. RNA folding, tertiary structures and 3D

1.2.1. *Secondary structures and RNA folding*

A simple and intuitive model is used to understand RNA folding. In RNA structures, an RNA chain folds on itself in a coil generating ~70% antiparallel double-stranded helices.

In the remainder of this book, we will refer to a double-stranded antiparallel helix, i.e. the canonical helix, as a helix. Antiparallelism is an important property because it allows the ends of the strands forming the helices to be connected with a minimum number of three nucleotides. The loops that close the helices are called apical. The continuity of a Watson–Crick helix is often broken by a sequence of nucleotides that form other structures called internal loops. While internal loops link several helices together, we speak of a junction with three, four or five helices. These elementary structures are called "hairpins".

The loops that close the helices may be larger or allow for the interlocking of other structures. The alternation of helices and loops defines the secondary structure as the alternation between helical and single-stranded regions (Figure 1.4). The formation of helices results from the stacking of bases enabled by the favorable conformations of the ribose-phosphate backbone and by the exclusion of solvent (water and ions). This phenomenon is promoted by enthalpy (ΔH) and then RNA compacts by releasing water molecules and ions to the solvent. Solvent disorder and thus the entropy (ΔS) of the system increase. The Gibbs free energy ($\Delta G = \Delta H - T\Delta S$) is therefore globally favorable (Woodson 2010).

Fundamentals of RNA and Ribozyme Structure 9

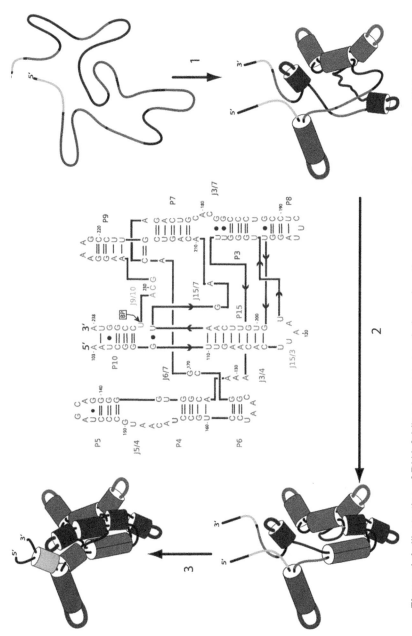

Figure 1.4. *Illustration of RNA folding on an example based on the lasso-capped ribozyme from Didymium iridis (Johansen and Vogt 1994; Nielsen et al. 2005) whose secondary structure is shown in the center*

COMMENTARY ON FIGURE 1.4.– *Helices are numbered Pi (paired) following the 5'–3' orientation (shown by arrows); junctions between helices are named Ji/j. The residues are numbered from 103 to 238. The motifs described in the text are found: the apical loops that close the helices, the internal loops (J5/4), the multiple junctions with three helices (P3/P8/P15), the pseudoknots (P3/P7 and P7/P15). The RNA structure is formed by the condensing into helices (colored cylinders) alternating with loops (thick lines) of the strand transcribed from DNA during transcription (1). Steps (2) and (3) illustrate the progression of folding and the gradual establishment of interactions that result in the appearance of new helices symbolized by cylinders and lead to the final, native structure.*

The remaining ~30% of nucleotides are generally represented in secondary structures as "orphan" residues. However, the majority of them also interact, which fits these residues into a given structural motif. These interactions can involve the three edges W, H, S of the nucleotides. Bases are thus assembled in pairs, triplets and quadruplets[2]. The constraints increase with the complexity of the interactions.

For example, guanines form quadruplets that stack to form telomeres (Figure 1.5) stabilized by potassium ions. The analysis of RNA structures contained in the PDB (Protein Data Bank)[3] allows the occurrences of each possible base pair to be counted according to the W, H or S edges involved (Stombaugh et al. 2009). This nomenclature developed by Leontis and Westhof associates a symbol with each edge in order to exhaustively describe all pairings of a secondary structure (Leontis and Westhof 2001). When single-bond hydrogen interactions exist, they always involve one edge of each involved base, which allows the pairing geometry to be characterized according to the nomenclature explained in Figure 1.5.

---

[2] A well-known example of a quadruplet is the assembly of four guanines together. These G-quadruplets stack up to form telomeres, which constitute the ends of chromosomes and are therefore essential to them. These quadruplets are also found in the RNAs being transcribed, which contain GGC repeats and are at the origin of Fragile X syndrome or possibly Charcot disease.

[3] Throughout this book, molecular structures are discussed. They are extracted from the open access database called the Protein Data Bank (https://www.rcsb.org). Historically, this database only contained protein structures. Each entity in this database is assigned a four-character code that enables download. These codes are indicated in this book.

Fundamentals of RNA and Ribozyme Structure 11

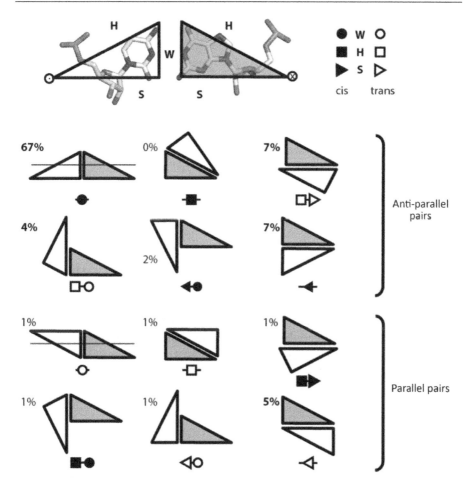

Figure 1.5. *Symbolizing nucleotides by rectangular triangles allows each edge to be characterized*

COMMENTARY ON FIGURE 1.5.– *The short edge, long edge and hypotenuse are assigned to the Watson–Crick, sugar and Hoogsteen edge, respectively. The top base pair materializes the orientation of the ribose-phosphate backbone using a vector perpendicular to the plane of the sheet, collinear with the helix axis. In a Watson–Crick cis base pair, the left strand ascends towards the observer's eye (who sees the head of the archer's arrow), while the right strand descends (the observer thus sees the tail of the archer's arrow). Twelve base pairs forming at least two hydrogen bonds are possible. Six pairings induce a parallel orientation of the strands and the other six induce an anti-parallel orientation. The percentages indicate the relative occurrences of each pairing in the PDB (Protein Data Bank) RNA structures (Stombaugh et al. 2009). A* cis-*pairing sees riboses located on the same edge with respect to an axis parallel to the hydrogen bonds between bases. We find that the major contribution is due to the* cis *W–W geometry, which is responsible for the formation of regular helices (67%). H–S,* cis *and* trans *S–S interactions come second with only 7%, 7% and 5%, respectively. The last significant contribution comes from* trans *H–W pairings. The other pairings contribute anecdotally. Thus, among the 12 possible pairings, five geometries are mainly observed.*

Certain combinations of base pairs and interactions generate specific motifs found recurrently in RNA structures, such as helices. The helix is only one of these motifs even though it is mainly in RNAs. An important motif, based on the helix structure, is the pseudoknot. But many other motifs exist that generate structural irregularities forming recognition platforms for other molecules such as metabolites (vitamins, sugar) or proteins in RNPs (ribonucleoprotein particles) such as the spliceosome or ribosome. Sections 1.2.2–1.2.6 describe some of them in order to illustrate the structural and functional diversity of RNAs.

1.2.2. *The pseudoknot*

The pseudoknot is the interaction between a region of the simple stranded RNA and the nucleotides of a loop connecting the two strands of a helix. The twist on the right edge of the helix prevents the single strand from passing into the loop which closes it. This feature gives its name to this motif, which is in fact only a "pseudoknot" in the topological sense (Pleij 1990; Westhof and Jaeger 1992), i.e. whatever deformation the RNA undergoes, the pseudoknot is maintained.

The interaction sees the establishment of Watson–Crick *cis* base pairs (Figure 1.6). The fact that one strand of the pseudoknot helix is composed of the nucleotides of a loop that can be quite short ($4 \leq n_{residue} \leq 12$) constrains how the base pairs at the terminals of the helices position with respect to each other. The relative positions of the two helices in the pseudoknot are thus imposed by the conformation of the loop. Pseudoknots are common in complex structure RNAs. However, they are not predictable from secondary structure prediction programs such as Mfold (Zuker 2003) or RNAfold (Lorenz et al. 2011) because their algorithm excludes the participation of the loop of one helix in the formation of another helix.

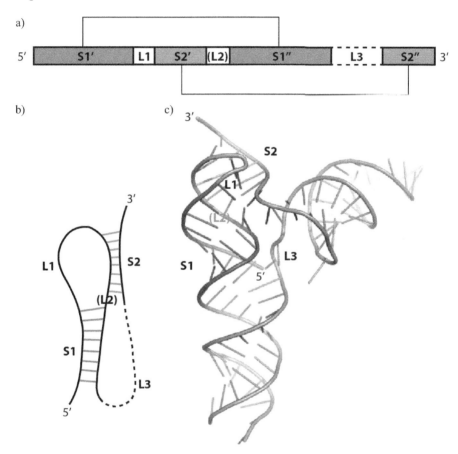

Figure 1.6. *The different representations of a pseudoknot*

COMMENTARY ON FIGURE 1.6.– *a) The elements that make up the pseudoknots can be shown in a linear sequence (symbolized here by a succession of boxes oriented from 5' to 3'). The S1' (S for stem) and S1" stems that constitute the first helix (blue) flank the S2' strand of the second helix (orange), while the second S2" strand is located at the 3' end. Three loops (L for loop) logically separate these four strands. b) Diagram of the transcription of the linear representation into a secondary structure. c) The three-dimensional representation shows the spatial organization of a pseudoknot in a global RNA structure. The structure of a riboswitch recognizing the queuosin precursor (pdb 4rzd (Liberman et al. 2015)) clearly shows the constraints on the L1, L2 and L3 loops. Because S1 and S2 are stacked on top of each other, L1 can be relatively short because it connects S1 to S2 in the deep groove. L2 is often non-existent. L3, on the other hand, can be very long and contains several tens or even thousands of nucleotides.*

Other algorithms, such as Kinefold (Xayaphoummine et al. 2003), which allows for transcriptional folding, must be used to analyze them, or other algorithms, such as ProbKnot (Bellaousov and Mathews 2010) or VSFold (Dawson et al. 2007), search for pseudoknots in a step that follows the secondary structure to be established (Jabbari et al. 2018). The most reliable method is identification by comparative analysis of nucleotide sequences and by biochemical mapping methods (Hajdin et al. 2013). But for this, it is necessary to have enough homologous RNA sequences to identify them.

The description of the topology of pseudoknots allows us to understand their importance in the global folding of an RNA. The single strand interacting with the loop can be located in the immediate vicinity of the stem-loop or, on the contrary, it can be very distant in terms of the number of nucleotides. The local structuring of an RNA can thus result from the condensing of regions that are close or very distant in sequence from each other. Whether RNA structural motifs are formed from local elements or between distant regions, their formation impacts the secondary structure by offering non-helical nucleotides the opportunity to form specific interactions.

1.2.3. *The loop E*

Initially characterized in prokaryotic 5S ribosomal RNA (Correll et al. 1997), this motif is also found in multiple copies in other ribosomal RNAs as well as in other RNAs such as group I and II autocatalytic introns.

Fundamentals of RNA and Ribozyme Structure   15

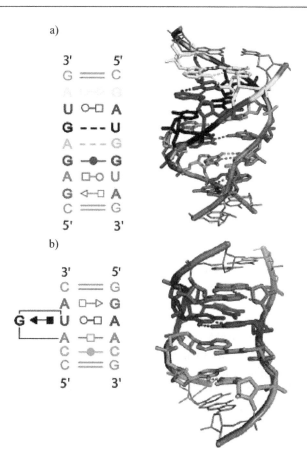

Figure 1.7. *Prokaryotic and eukaryotic loops E*

COMMENTARY ON FIGURE 1.7.– *a) The loop E of prokaryotic 5S RNA consists of the sequence of two characteristic pairings (yellow and blue). Additional specific pairings are added to this basic motif (pdb 354D (Correll et al. 1997)). The two characteristic pairings repeat at the other end of the loop following a second-order symmetry (red and green). At the center of the loop, two R–R pairings and one R–Y pairing take place (purple, orange, black). For steric reasons, these pairings do not rigorously adopt a Watson–Crick geometry but are bifurcated. b) The loop E of the sarcin/ricin of eukaryotic 28S RNA (pdb 480D (Correll et al. 1999)) is different from that of prokaryotic 5S RNA but identical to the corresponding region of 23S RNA. It has no symmetry and consists of a 3-base pair motif plus an S-turn stabilized by the interaction of the sugar edge of the unpaired guanine with the Hoogsteen edge of the uracil (blue).*

The loop E is based on a precise sequence of non-Watson–Crick base pairs that gives it its structural signature (Figure 1.7). It has been identified also in prokaryotic 23S and eukaryotic 28S ribosomal RNAs (Correll et al. 1999) in slightly different but functionally equivalent forms. It is able to interact with proteins such as the toxins sarcin and ricin that cleave the RNA chain resulting in the inability of the rRNA to bind the translation elongation factors Tu and G (Correll et al. 1999). This motif retains a linear helical axis of the helix and thus adopts a cylindrical shape.

On the contrary, the grooves are modified in width and depth giving them the structural properties, meaning the proteins mentioned above can be recognized.

However, like all RNA structural motifs, the loop E is capable of recognizing other RNAs as in the group I intron of *Tetrahymena* where an extension adopts this structure that stabilizes the catalytic domain (Waldsich et al. 2002). This motif is also found in the hairpin ribozyme where it forces the substrate strand to adopt a conformation that promotes the cleavage reaction (Rupert et al. 2002).

1.2.4. *The k-turn*

Unlike the loop E, the interplay of nucleotide interactions in the k-turn (or kink-turn) (Figure 1.8) induces the formation of a kink and thus enables the flanking RNA helices to point in different directions by forming an angle of about 60° between their helical axes; this means the shallow grooves face each other (Klein et al. 2001).

These motifs allow RNAs to break out of the linearity imposed by the helical structure itself and thus to form complex structures. The k-turn also constitutes an interaction platform for certain proteins.

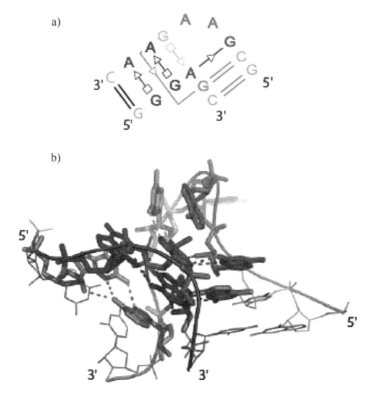

Figure 1.8. *The k-turn is an asymmetric inner loop that interrupts the helical continuity by creating a bend of about 60° that places the shallow grooves of the two helices face to face*

COMMENTARY ON FIGURE 1.8.– *In this motif, the thickness of the nucleotides' atomic bonds involved in the specific interactions has been increased. The motif is stabilized by a triple interaction formed between the GA base pair (blue) and the GC (gray). a) The secondary structure on which the non-canonical base pairs are represented in the Leontis–Westhof nomenclature is deduced from b) the structure of the nucleotide 77 region of the 50S ribosomal subunit of Haloarcula marismortui (pdb 1jj2 (Klein et al. 2001)).*

The most studied example is the 15.5 kDa protein[4] that recognizes k-turns present in spliceosomal RNAs responsible for pre-mRNA splicing (Vidovic et al. 2000; Szewczak et al. 2005) and in the C/D boxes of small nucleolar RNAs (snRNAs) which guide the enzymes of chemical modifications of ribosomal RNAs (Baird et al. 2012). Ribosomal RNAs are also provided with them each time, the key being the recognition by a ribosomal protein specific to the region of the ribosomal RNA considered.

In fact, it was in the structure of the 50S ribosomal subunit that this motif was initially described (Klein et al. 2001). RNase P (Cho et al. 2010; Meyer et al. 2012), as well as several riboswitches, exhibit this motif (Montange and Batey 2008; Lee et al. 2010).

1.2.5. *Tetra-loop receptors*

In addition to local organization patterns such as the loops and k-turns, there are also long-distance organization patterns that are distinct from pseudoknots. The archetypal of these motifs is the tetra-loop receptor, which has two modes in which either a regular helix or a highly irregular non-canonical helix enables specific recognition of the loop.

Two families of apical tetra-loops are particularly stable. The first family adopts a 5'-UNCG-3' consensus sequence. Typically, these loops do not interact with any other region of the RNA although recent work indicates that they may do otherwise which induces a conformational change (D'Ascenzo et al. 2017, 2018; Meyer et al. 2019).

In contrast, different interaction modes of the second family of 5'-GNRA-3' consensus sequence loops were obtained as early as the resolution of the first crystallographic structures of ribozymes (Pley et al. 1994a; Cate et al. 1996a). Their helix recognition modes had previously been deduced by molecular modeling approaches (Michel and Westhof 1990) based on biochemical data (Costa and Michel 1995). The first mode, which is simple in its elegance, sees two-loop adenines recognize the shallow groove of a helix with at least two contiguous G=C base pairs (Figure 1.9(a)). This first pattern was first observed in the hammerhead ribozyme crystal stack (Pley

---

4 The kilo Dalton (kDa) corresponds to 1,000 times the mass of a hydrogen atom. This unit is used in biochemistry to characterize the mass of proteins and biological polymers in general.

et al. 1994a). The second, more complex mode is observed in an autonomous structure domain of the group I intron of *Tetrahymena thermophila*. Three adenines of a 5'-GAAA-3' loop interact with a helix whose irregularity flows from the concatenation of a G=C base pair, an A(H–W)U *trans* pairing and finally a pair of consecutive adenines of the same strand forming an -A(S–H)A platform (Cate et al. 1996a, 1996b). This receptor is perfectly adapted to the recognition of the loop. The G=C base pair recognizes the 3' adenine of the loop (A2). The second adenine (A1) interacts with the sugar edge of the receptor's uracil, while the first adenine (A0) forms a *trans* Watson–Crick base pair with the receptor's adenine whose Hoogsteen edge is already interacting with the uracil's Watson–Crick edge. This first adenine is also stacked on the AA platform, conferring additional stability to the assembly (Figure 1.9(b)).

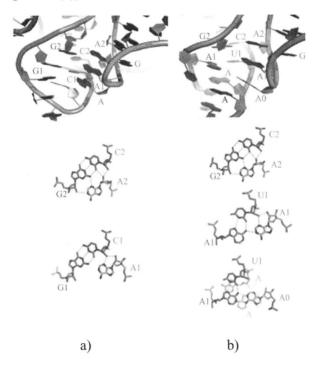

a)   b)

Figure 1.9. *The different modes of interaction between a GNRA loop and its two main types of receptors*

COMMENTARY ON FIGURE 1.9.– *a) Interaction between a GNRA loop and the shallow groove of a helix in the crystal stack of the hammerhead ribozyme structure (Pley et al. 1994a). The two constituent base triplets of the interaction are detailed below the overall motif representation (top panel). The A2 nucleotide interacts with the nucleotide of the nearest helix; the right-hand twist of the RNA helix induces a shift of A1 towards the nucleotide of the opposite strand. In this motif, the same nucleotide (adenosine) interacts with the shallow groove of a G=C pair in two different ways. b) A GNRA loop frequently interacts with a specific receptor. This receptor is centered around a U(W-H)A base pair with which A1 interacts. This UA pair is framed by a G=C pair and an A–A platform on which A0 rests (Cate et al. 1996a, 1996b). The G2-C2/A2 triplet is the same in cases (a) and (b). The context in which the tetraloop–receptor interaction is observed is the P4–P6 intron domain of Tetrahymena thermophila (see Figure 1.10(a)).*

1.2.6. *The A-minor pattern*

The irregular receptor of GNRA loops is observed much less often than a succession of two or three G=C pairs. G=C pairs, which are more stable than A–U pairs, are preferred points for making tertiary contacts. Similarly, a sequence bias exists on loops (apical or internal) where adenines are individually more frequent than other nucleotides. Because their aromatic rings have a large surface area that is relatively isomorphic with an exocyclic amine (the N6 group), adenines tend to spontaneously stack with each other. These stacks form blocks whose accessible edges can interact with the shallow groove of a helix according to S>W>>H preferences. These arrangements are identical to the interactions observed between the two adenines of a tetra-loop and two G=C base pairs described in section 1.2.5. This relatively common situation provides a significant free energy gain for RNA folding.

The interaction between the shallow groove of a helix composed of G=C pairs and single-stranded adenines from a loop composes a versatile tertiary structure motif called A-minor (Doherty et al. 2001; Nissen et al. 2001). In this motif, one, two or three consecutive adenines interact with two to three successive base pairs of a helix in characteristic recognition modes

(Figure 1.10). In the first two modes, free adenines interact primarily with their S edge. The adenine slides towards the residue to which it is antiparallel so its W edge interacts with the O2' in the third mode. The O2' – which fundamentally distinguishes RNA from DNA – is again of particular importance here and promotes structural diversity in RNA. The preponderance of the O2' group necessitates another remark. As it is present in every nucleotide, as well as N3, or O2, its equivalent in pyrimidines (Figures 1.1 and 1.2), A-minors are in essence undetectable by comparative sequence analysis unless other information points to them. These contacts were originally observed in the crystallographic structures of ribosomal subunits.

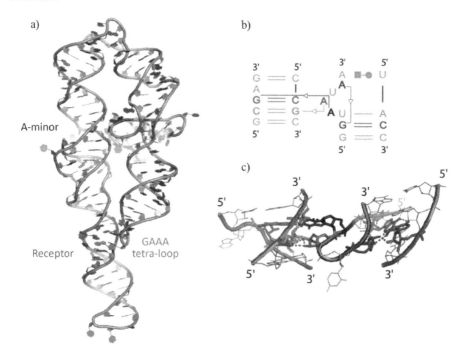

Figure 1.10. *An example of an A-minor interaction in the P4–P6 domain of the* Tetrahymena thermophila *intron where this interaction was first studied by site-directed mutagenesis (Doherty et al. 2001)*

COMMENTARY ON FIGURE 1.10.– *a) A global view of the P4–P6 region of the group I intron of* Tetrahymena thermophila *(PDB 1gid (Cate et al.*

1996a)) visualizes the establishment of the A-minor interactions (right side) between the single-stranded adenines and the G=C pairs of the helix. (b) The detailed secondary structure of the interaction highlights the contacts between the S edges of the involved nucleotides according to the Leontis–Westhof nomenclature. c) The three-dimensional structure shows how the single-stranded loop folds back on itself so as to position the adenines favorably to interact with the shallow groove of the helix in front. The two stacked A-minor structure exactly as in Figure 1.9(a). The positions of the tetra-loop and its receptor are also shown in the color code corresponding to that in Figure 1.10(b).

These versatile A-minors have a relatively low individual energy contribution, but their number contributes greatly to the establishment of the tertiary structure. This motif is therefore extremely common in RNA structures because it promotes the flexibility that allows the tertiary structure to adopt the different native states, i.e. the structures in which the RNA is able to perform its biological function. The low individual energy contribution of this motif largely favors RNA dynamics and thus plays a prominent role in biological mechanisms. Their involvement in the process of messenger RNA decoding via the ribosome perfectly illustrates this idea (Lescoute and Westhof 2006; Demeshkina et al. 2012, 2013).

The small ribosomal subunit is the reading head that iteratively decodes the codons written on the mRNA using the anti-codon loop at the decoding site. In a simplified way, if the tRNA corresponds to the amino acid indicated by the codon sequence, then a complementary anti-parallel Watson–Crick helix is formed between the codon and the anticodon. The decoding site has adenines in a loop of the 3' domain that recognize the shallow groove of this helix (Figure 1.11). If this helix is regular, the decoding site adopts a conformation that establishes A-minor contacts. Such a conformation then indicates to the ribosome that the recognized tRNA corresponds to the codon being decoded. On the contrary, if the codon–anticodon interaction is not complementary, the information conveyed to the ribosome is to reject the tRNA in place (for a more complete overview of this mechanism, see Rozov et al. (2016b)). Whether or not the tRNA matches the mRNA codon, this correction or proofreading step uses energy because in both cases the Tu elongation factor that brings the aminoacylated tRNAs to the ribosome hydrolyzes a GTP molecule to leave the ribosome. In the case of codon–anticodon correspondence, the transpeptidylation reaction takes place.

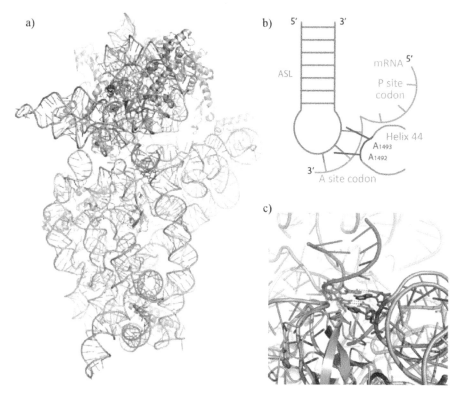

Figure 1.11. *The small ribosomal subunit detects codon–anticodon complementarity using the A-minor motif*

COMMENTARY ON FIGURE 1.11.– *a) The small ribosomal subunit is the messenger RNA (mRNA) reading head. It is the center where the interaction between the codons and anticodons of tRNAs carrying the amino acids (aa) to be polymerized occurs during the elongation phase of protein translation. The subunit's A site is the decoding site for this interaction. The P site is the site to which the tRNA carrying the synthesizing peptide binds. Translocation of the mRNA shifts the codon from the A site to the P site and thus brings a new codon to the A site to which a new tRNA can fit. These steps allow for a polymerization rate of approximately 20 aa per second. When the codon–anticodon helix forms, adenines 1492 and 1493 (red) of helix 44 (gray) of 16S RNA change conformation and form A-minor interactions with the shallow groove of the helix formed between the mRNA (orange) and tRNA (cyan) at the A site (B, C). If the conformation is correct, it is transduced to the ribosome as a positive signal to integrate the amino acid into the synthesizing peptide chain. On the contrary, if the groove does not adopt a helical conformation, the A-minors are not formed and the conformation change does not take place, which induces the release of the tRNA. The amino acid is not incorporated. b) Schematic representation of the codon–anticodon interaction at the A-site. c) Three-dimensional view of the codon–anticodon interaction at the A-site (pdb 1ibm (Ogle et al. 2001)).*

In the simplified folding model described above, the secondary structure and local structural motifs such *as k-turns* or the loop E orient in space the various structured elements of the RNA, which induces the formation of tertiary interactions of the native structure. A final step can then occur locally when a tertiary interaction has sufficient energetic contribution to allow a modification of the secondary structure to a less locally stable but globally favorable state (Koculi et al. 2012).

1.2.7. Comparative analysis of sequences

Sequence information allows the proposal of refined secondary structure models by comparing the initial sequence to homologous RNA sequences identified in other organisms. As stated above, the most represented motif in RNA structures, the A-helix, results from the formation of Watson–Crick base pairs that are isosteric to each other. Therefore, sequence variations that preserve Watson–Crick *cis-pairings* do not affect the helical structure. When homologous functional RNAs are identified in different organisms, their three-dimensional structures are largely related. This means that most helices will occupy similar positions relative to each other in space and will often be of the same length. If sequences are important for the function of the molecule, they will also be more conserved than the sequences of the helices. It is therefore possible to "align" the sequences by creating an array where each nucleotide of the initial sequence is stacked with the equivalent nucleotides of the other sequences (Figure 1.12). The rows of the table represent an RNA of a given species, and the columns represent the identities of the nucleotides at each position. Specific software, Jalview and Assemble (Waterhouse et al. 2009; Jossinet et al. 2010), allow editing of such alignments. First, the conserved nucleotides are collected in the same columns. Then, the helices are identified by locating columns that vary non-independently (covariations). Insertions and deletions reflect the structural differences often observed between homologous RNAs, as they perform identical functions but in specifically adapted biological contexts. To identify helices, secondary structure folding programs such as Mfold (Zuker 1989) and RNAfold (Hofacker et al. 2002) can be used, but it is preferable to use programs capable of predicting pseudoknots such as Kinefold (Xayaphoummine et al. 2003) or the Rivas–Eddy (1999) or Reeder–Giegerich (2004) algorithms.

Programs that fold the secondary structure within a sequence alignment have been written. Some align the individual secondary structures of each RNA; others are based on the Sankoff algorithm (1985) that simultaneously folds and aligns sequences. Probabilistic models (CM, covariance models (Weinberg and Ruzzo 2004, 2006; Nawrocki and Eddy 2013)) can also be inferred from initial sequence alignments to search genomes for homologous candidates. However, we limit ourselves here to referencing key works in order to help the reader who wishes to go further, as the topic of comparative sequence analysis and associated algorithms alone would require a dedicated book.

In addition, to cope with the exponential increase in the number of RNA sequences due to the contribution of high-throughput sequencing methods (NGS: Next Generation Sequencing), databases dedicated to sequence alignments by RNA families have been created since the 2000s. One of the best known is the RFam database (Gardner et al. 2011; Kalvari et al. 2021), which presents the different known RNA families linked to Wikipedia and provides representative sequence alignments that generate statistical models that are used to identify homologous RNAs in genomes. However, the increase in the number of sequences also induces the increase in the number of specific databases that have different formats and languages. Interfacing these heterogeneous data is difficult. In order to overcome this problem, a search engine capable of probing the majority of RNA databases has been designed on the model of a central station, RNAcentral[5]. RNAcentral provides the maximum of up-to-date information on a given RNA and is therefore a good starting point to become interested in a topic related to non-coding RNAs.

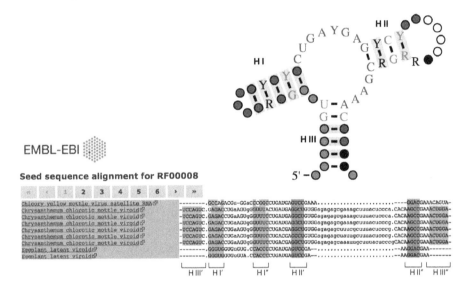

Figure 1.12. *The RFam database (Kalvari et al. 2021)[6] of the European Bioinformatics Institute (EBI) hosted by the European Molecular Biology Laboratory (EMBL) collects published RNA sequences and offers many services for studying their folding*

COMMENTARY ON FIGURE 1.12.– *The figure constructed from the information given by RFam shows how an alignment of a few hammerhead ribozyme sequences allows us to deduce the pairings that define the Watson–Crick helices (colored blocks). In this example, the HIII sequence (formed by the interaction between the 5' HIII' and 3' HIII" strands) is conserved, but covariations are observed between the HI and HII strands. This example also shows how the conserved sequence blocks allow for helix shimming. Insertions (shown in small print) are visible in the HI and HII*

---

5 See: https://rnacentral.org/help.
6 See: http://rfam.xfam.org/.

loop. The complete alignment, of which only a fraction is shown here, allows us to infer a consensus secondary structure with the R-scape method (Rivas et al. 2017). The conserved nucleotides are indicated in full. Nucleotides that vary are indicated by disks whose color indicates the conservation rate (white: 50%; gray: 75%; black: 90%; red: 97%).

2

Ribozymes and the "Central Dogma" of Molecular Biology

2.1. The discovery of RNA catalysis and the central dogma of molecular biology

The discovery of the catalytic properties of RNAs was rewarded in 1989 with the Nobel Prize in Chemistry awarded to Thomas Cech and Sidney Altman. Catalytic RNAs, since then called ribozymes for ribo(nucleic-en)zymes, are now also considered as enzymes. At that time, RNAs were considered as simple messengers between the depository of genetic information, DNA, and the products of the decoding of this information, the proteins which alone, it was believed, ensured biological functions. The discovery of the helical structure of DNA in 1953 by James Watson and Francis Crick (1953) is the basis of the "central dogma" of molecular biology in which DNA and proteins are the major players (Figure 2.1). From the 1970s, the discovery of mechanisms associated with RNA metabolism such as reverse transcription, splicing and RNA editing refined the "central dogma" and placed RNA at the center of cellular processes. It has also been known for a long time that many families of viruses – herpes, hepatitis, coronavirus, human immunodeficiency virus, to name a few that pose enormous human health problems – have a genome carried by RNA that is both a carrier and messenger of genetic information (Figure 2.1).

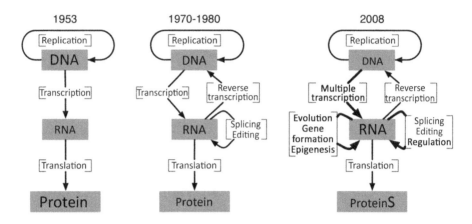

Figure 2.1. *The central "dogma" of molecular biology and its evolution since its initial formulation in 1953. In addition to the appearance of new arrows indicating the discovery of original cellular phenomena such as splicing, reverse transcription or RNA-dependent transcription, the thickness of the arrows is proportional to the relative importance of the different phenomena*

It was not until the early 1980s that the catalytic properties of RNA were demonstrated and therefore the first ribozymes identified. These historic ribozymes are involved in two important cellular processes, first of all the splicing of RNA precursors, i.e. the ligation of the coding regions which are the exons following the removal of the non-coding regions that are introns (Figure 2.2). Here, the notion of coding exons is understood as the RNA precursor elements that are expressed of course as proteins as well as RNAs, such as ribosomal and transfer RNAs (autocatalytic introns often contain regions that code for proteins such as reverse transcriptases (Belfort and Lambowitz 2019) or endonucleases that recognize specific cleavage sites (Haugen and Bhattacharya 2004; Haugen et al. 2004; Hafez and Hausner 2012)). In the context of ribozymes, this mechanism relies on the action of an autocatalytic intron that, by folding back on itself, has the ability to successively cut the bonds that connect it to the two exons that surround it while ligating them. The second case concerns the maturation of transfer RNAs (tRNAs) by an enzyme whose nature had not yet been characterized (Figure 2.3). This enzyme, a ribonucleoprotein, has a protein part but the RNA part carries the catalytic activity. These two RNAs catalyze transesterification reactions that rely on the attack of a phosphate group (PO_4^-) by a hydroxyl function (-OH) (Figure 2.4). This hydroxyl group can come from a nucleotide (2' or 3' positions) in the case of autocatalytic introns or from a water molecule (H-OH) in the case of RNase P.

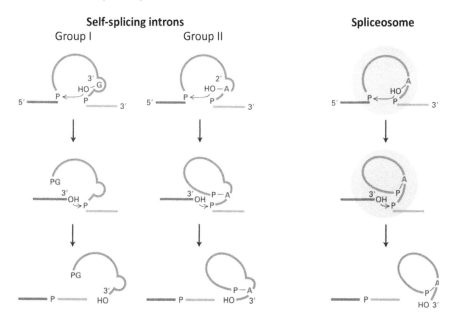

Figure 2.2. *Splicing by group I and II introns and the spliceosome (from W.H. Freeman, Molecular Cell Biology, 7th edition (Berk et al. 2016)). For a color version of this figure, see www.iste.co.uk/masquida/ribozymes.zip*

COMMENTARY ON FIGURE 2.2.– *Intron splicing is performed in two transesterification steps that cleave the 5' and 3' terminals of the intron (gray) in sequence and ligate the exons (pink). The group I and II introns are autocatalytic. Despite their profound structural differences, their catalytic cycles are similar. Catalysis of group I introns relies on the activity of an exogenous guanosine (e.g. a GTP molecule, naturally abundant in the cell). In the first step, a region of the intron places this GTP molecule so that one of the hydroxyl groups cuts the link between the 5' exon and the intron. The* cut exon in turn becomes nucleophilic due to a conformational change. The reaction ligates the two exons and cuts the intron of exon 3'. This is the second step. In the case of group II introns, the nucleotide that carries the nucleophile in the first step is an adenosine belonging to a particular domain of the intron. The case of spliceosome is more complex because this system contains five different RNAs and many associated proteins. Nevertheless, it is also a nucleotide carried by an RNA that serves as a nucleophile in the first step. The spliceosome is therefore also a ribozyme.

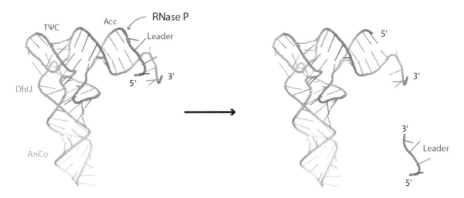

Figure 2.3. *The first step of maturation of tRNA by RNase P allows the leader 5' end of pre-tRNAs to be cut. For a color version of this figure, see www.iste.co.uk/masquida/ribozymes.zip*

COMMENTARY ON FIGURE 2.3.– *The position of the cleavage is materialized by a red arrow. The tRNAs are composed of four domains each with a specific name, the acceptor arm (Acc) that carries the amino acid to the ribosome, the thymine (TΨC), dihydrouridine (DhU) and anticodon (AnCo) stem-loops. The latter pairs with the corresponding codon of the mRNA on the ribosome that verifies the compliance of the interaction (see Chapter 1). The representation of the tRNA precursor is developed from the structure of yeast RNase P in complex with the phenylalanine tRNA precursor (pdb: 6ah3 (Lan et al. 2018)).*

Ribozymes and the "Central Dogma" of Molecular Biology 33

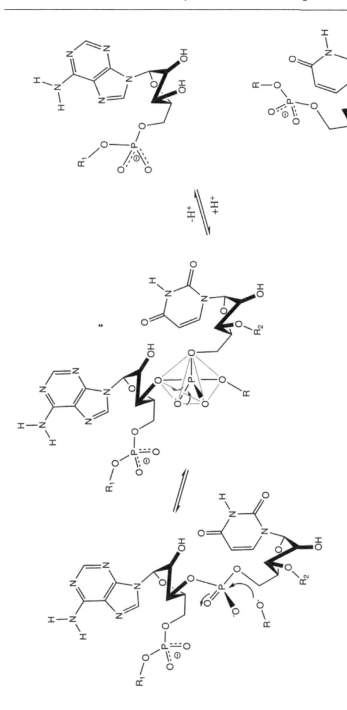

Figure 2.4. *The transesterification reaction in group I and II introns and RNase P. For a color version of this figure, see www.iste.co.uk/masquida/ribozymes.zip*

COMMENTARY ON FIGURE 2.4.– *The activated nucleophilic species (R–O⁻) which is either a hydroxide ion or a 2' or 3' deprotonated nucleotide attacks the targeted phosphate group forming an oxyphosphorane transition state (‡) adopting a trigonal bipyramid geometry (pink lines) carrying two negative charges. The transition state hydrolyses spontaneously forming 3' hydroxyl and 5' monophosphate ends. Obtaining these products is favored by the architecture of group I and II introns and RNase P. Autonucleolytic ribozymes give different products. The negative charge of the phosphate group is neutralized by a cation like Mg^{2+} which makes the phosphate more electrophilic.*

2.2. In search of the primordial polymerase

Transesterifications are also observed in DNA during genetic recombination steps. In addition to their capacity to store genetic information, RNAs can have their linear structure manipulated by fragment insertion or deletion. These properties are necessary and sufficient to develop prebiotic molecules capable of storing information while being self-replicating. These RNAs should therefore have been able to polymerize nucleotides while using their own sequence as a template. Such RNAs do not exist or no longer exist in nature, just as 90% of the living species that populated the Earth have disappeared. But teams of researchers have obtained in vitro RNAs capable of polymerizing sequences of tens of nucleotides (Lawrence and Bartel 2005; Shechner and Bartel 2011; Cojocaru and Unrau 2021), including copying themselves (Horning and Joyce 2016; Tjhung et al. 2020). These results offer evidence that an RNA world is not just a concept as several prominent researchers had imagined starting in the 1960s (Woese 1967; Crick 1968; Orgel 1968; Gilbert 1986) by noting how the process of protein synthesis is subservient to interactions between messenger and transfer RNAs, guided by ribosomal RNAs. These properties could have gradually led to the emergence of living systems. The discoverers of ribozymes therefore hypothesized that an RNA world could have existed before the first living cells. The arguments in favor of this hypothesis are detailed in several reference books for the scientific community (Gesteland and Atkins 1993; Gesteland et al. 1999, 2006).

From the 1980s onwards, discoveries highlighting the role of RNAs as a major player in molecular and cellular biological functions have followed one another (Figure 2.1). According to the RNA world hypothesis, RNAs could be at the origin of life and would have provided the first support for genetic information before being retro-transcribed during evolution into genes on a DNA material. This transition from RNA to DNA would have stabilized genetic heritage and improved the fidelity of its transmission over the generations. An element reinforcing this idea comes from the observation that virus genomes in RNA form are limited to a size of about 30 kb[1], which implies that beyond this size maintaining the integrity of the genome is compromised (Ferron et al. 2021). In the living world as we know it, the epigenetic roles of RNAs (non-heritable mechanisms modifying gene expression) and in the regulation of genetic expression during the transcription, post-transcription and protein translation steps would indeed be the remnants of an RNA world. RNAs therefore play a major role in the living world in the formation and evolution of genes by providing a genetic material that is more malleable before being selectively "fixed" on a DNA material.

---

1 The lengths of genomes, plasmids and nucleic acids in general can be expressed in kilobases (kb), which correspond to the thousands of nucleotides they contain.

3

The Discovery of Ribozymes

We will now look back at how these discoveries that changed biology in the 1980s were made. At that time, research on the mechanisms of protein translation was dominating the biologist community, which wanted to understand how the ribosome worked. The research that was going to highlight the catalytic activity of RNAs therefore seemed anecdotal.

In this context, two American laboratories simultaneously isolated two distinct RNAs capable of specifically cutting an RNA chain (transesterification). These RNAs, and no doubt others, were thus endowed with a nuclease activity. The knowledge of the time on biological catalysis, which was thought to be based only on proteins, was profoundly altered.

The first laboratory, led by Thomas Cech at the University of Boulder (Colorado, USA), demonstrated that an intron interrupting a ribosomal RNA gene of the protist *Tetrahymena thermophila* spontaneously cut itself and ligated the 5' and 3' fragments, generating a mature ribosomal RNA.

The second laboratory, led by Sidney Altman at Yale University, showed that the RNA subunit of RNase P, an enzyme responsible for tRNA maturation, was catalytic. These discoveries, in connection with the global effort to understand translation, earned these two researchers the 1989 Nobel Prize in Chemistry.

3.1. The discovery of catalysis by autocatalytic introns

Initially, as recounted by Thomas Cech in his speech at the occasion of his Nobel Prize award (Cech 1989), his laboratory was studying the transcription of ribosomal DNAs from *Tetrahymena thermophila*[1], an organism whose ribosomal DNAs (rDNAs) located on autonomous chromosomes are specifically amplified in the nuclei (Orias et al. 2011). The goal was to isolate rDNAs carrying transcription complexes in order to identify the proteins involved. Nucleotide sequences did indicate the presence of a DNA region of approximately 400 nt interrupting the 26S ribosomal RNA gene, but this was of secondary interest because this type of genetic architecture had already been described. By incubating extracts of cell nuclei containing amplified rDNAs and transcription enzymes in the presence of nucleotide triphosphates (NTPs), Thomas Cech and his team detected RNAs synthesized in vitro. These experiments revealed an RNA fragment whose sequence matched that of the 400 nt interruption (Cech et al. 1981).

It became interesting to understand the mechanism by which the mature 26S RNA precursor is stripped of this extension before the ribosome is formed. At this stage, Thomas Cech was convinced that a protein enzyme was responsible for the appearance of this 400 nt RNA[2] since the transcripts were carried out in nuclear extracts of *Tetrahymena thermophila* which a priori contained all the active nuclear proteins. It was therefore necessary to find conditions allowing the production of RNAs in vitro while simultaneously inhibiting splicing, and then to expose these precursor RNAs to different fractions of nuclear extracts in order to identify those containing the "mystery" enzyme responsible for the cutting of the 400 nt extension. The experiment worked on the first attempt, but Thomas Cech noted that even the control sample, which contained only RNA and no protein, showed an RNA of 400 nt corresponding to the maturation product.

---

[1] *Tetrahymena thermophila* is a ciliated protozoan. It is a unicellular eukaryote close neighbor in the tree of life of the group of Apicomplexes to which belongs *Plasmodium falciparum*, well known to cause malaria.
[2] The length of a nucleic acid can be expressed in kb and also in nucleotides (nt).

A detailed analysis of the 5' sequence led to an understanding that it included an additional guanosine (G) in 5', demonstrating the need for a chemical addition reaction (Zaug and Cech 1986a, 1986b; Zaug et al. 1986).

The final experiment seemed to be straightforward. It was sufficient to verify whether a 400 nt radioactive RNA appeared following incubation of pre-ribosomal RNAs in a solution containing guanosine labeled with a radioactive phosphorus isotope (^{32}P) commonly used in molecular biology, the other NTPs being unlabeled.

This positive result in the absence of protein finally demonstrated that RNA was catalytic. It was still necessary to identify the details of the process (the first reaction is the one represented in Figure 2.4).

At the same time, a French research group, of which François Michel was a member at the CNRS in Gif-sur-Yvette, identified introns in the mitochondrial DNA of the yeast *Saccharomyces cerevisiae* with a structure close to that of the *Tetrahymena* intron, as well as a second category of introns that would later prove to be capable of catalyzing their own excision (Michel et al. 1982; Michel and Dujon 1983). The secondary structures of these different introns clearly indicated the existence of two classes that led to the denomination distinguishing group I introns and group II introns.

Nevertheless, these two classes of introns have in common the ability to catalyze two successive transesterification reactions to result in the ligation of exons.

In the case of group II introns, it is the 2' hydroxyl group of an adenosine domain located near the 3' end of the intron that provides the first nucleophile and enables the formation of a branched nucleotide called "lasso" or lariat[3] (Figure 2.2).

---

3 The name lasso comes from the fact that the attack by the 2' hydroxyl group of a nucleotide whose 3' hydroxyl is already linked to the RNA chain results in an RNA whose nucleotide presents all its hydroxyl groups in esterified form. This RNA is therefore no longer a simple linear chain. However, the node of the lasso is not sliding, which gives this term a slight ambiguity.

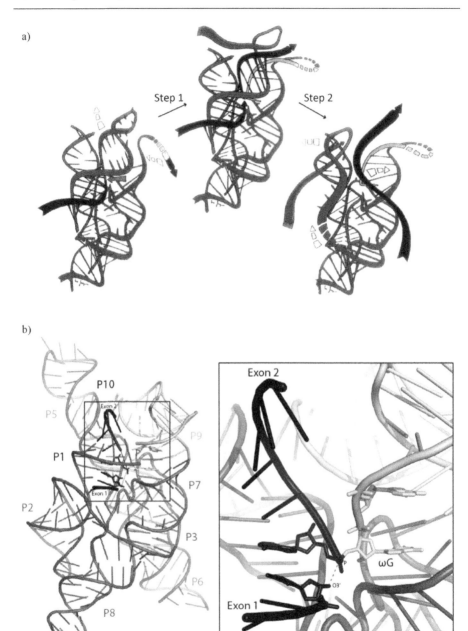

Figure 3.1. *Cleavage mechanism by group I introns. For a color version of this figure, see www.iste.co.uk/masquida/ribozymes.zip*

COMMENTARY ON FIGURE 3.1.– a) *Model of the catalytic core of group I introns. In step 1, the GTP molecule symbolized by the orange arrow cuts the bond between the 5' exon (black) and the intron whose substrate elements are colored red and the catalytic core blue. A conformational change (symbolized by the upward pointing hatched arrow) unwinds the substrate helix and allows the 3' end of the intron (yellow) to move into the catalytic site, dragging the 3' exon (black) with it. The 5' exon (black arrow) serves as a nucleophile during the second step which cleaves the link between the 3' end of the intron (yellow) and the 3' exon and the simultaneous ligation of the exons. The intron is then released. The drawings in this image were made by Dr. Luc Jaeger and published in his thesis (Jaeger 1993). b) The crystallographic structure of the group I intron (pdb: 1u6b) present in the tRNA^{Ile} gene of the bacterium* Azoarcus *(Adams et al. 2004a, 2004b) indicates that the predictions made by François Michel and Éric Westhof concerning the architecture of the catalytic site were correct. On the left, the overall structure of the intron in colors identical to those of the drawings in (a) enables the architectural similarities with the model published 14 years earlier to be understood. The A-form helices are indicated from P1 to P10. P1 and P10 result from pairing between the 3' ends of exon 1 and 5' ends of exon 2 (black) with the intron guide sequence (red), respectively. The inset on the right shows how the 3' hydroxyl of exon 1 is in position to attack the phosphate (P) moiety located between the last nucleotide of the intron (ωG, yellow) and the first nucleotide of exon 2 (red hatched line). At the end of the reaction, the bond between the intron and exon 2 will be broken and simultaneously both exons will be ligated. The intron sequences forming P1 and P10 have been omitted for simplification.*

In order to elucidate the RNA structures that allow the succession of splicing transesterification reactions, François Michel had the idea of aligning the sequences (see the introductory chapter on RNA structure) of several copies of group I introns in order to identify the conserved elements, on the one hand, and the helices, on the other, by covariation analysis, based on the principle that if these introns carried out the same reactions, their structures must present homologies. This is how he identified the two distinct classes of distinct introns. The elements P1–P10 of the catalytic core of group I introns (Michel et al. 1990) were assembled in a three-dimensional structural model in 1990 with the help of Éric Westhof (Michel and Westhof 1990). This model constituted for about 15 years the only structural reference on this subject and allowed a model of the sequence of the two catalytic steps to be proposed involving three specific conformers (Figure 3.1). At the beginning of the 2000s, the crystallographic structures published (Adams et al. 2004b; Guo et al. 2004; Golden et al. 2005) generally confirmed the architecture proposed by Michel and Westhof, but introduced new elements on the details of the interactions between nucleotides and on the catalytic aspects. It was also in François Michel's group that Maria Costa and her colleagues generated the first crystallographic structure of group II intron presenting the lasso resulting from the first stage of splicing (Costa et al. 2016), although group II intron structures not lassoing in the first step were published by Anna-Maria Pyle's group at Yale University in the meantime (Toor et al. 2008a, 2008b).

3.2. The discovery of RNA catalysis of RNase P

In the late 1970s, Sidney Altman (at the MRC Laboratory of Molecular Biology in Cambridge and later at Yale University) also highlighted the catalytic properties of RNAs in a biological process related to the maturation of transfer RNAs (tRNAs). In order to understand the behavior of mutant tRNAs induced by acridine, Sidney Altman purified total RNAs from the enterobacterium *Escherichia coli* (Altman 1971) and demonstrated the existence of tRNA transcripts whose ends are longer than those of the tRNAs used by ribosomes. Acridines are aromatic polycyclic compounds that intercalate between the base pairs of nucleic acid double helices, thereby disrupting the mechanisms of replication and transcription (Ferguson and Denny 1991). In all likelihood, these are tRNA precursors. Remarkably, these mutant transcripts had an increased life span compared to wild-type tRNA transcripts. This feature made them detectable in assays in contrast to wild-type intermediate transcripts, which rapidly disappeared.

As recounted by Sydney Altman in his Nobel Lecture (Altman 1989), incubation of these mutant transcripts in an *E. coli* cell extract sees the rapid production of tRNAs without additional 5' and 3' ends. The 5'-end cleavage enzymatic activity is isolated in bacterial extract fractions (Altman and Smith 1971). The specificity of this reaction for tRNA and the obtaining of 5' phosphate and 3' hydroxyl ends indicate that the enzyme named RNase P was different from other nucleases known at that time, which motivated researchers to study it with attention (Robertson et al. 1972). However, it was only after several years that the association between a small protein subunit and a 400 nt RNA necessary for the activity constituting the enzyme was understood (Stark et al. 1978). And it would take a few more years and the development of in vitro transcription techniques to characterize the nuclease activity carried by the RNA (Guerrier-Takada et al. 1983).

Initially, it already seemed surprising that the RNA subunit would be required for activity, but as in the case of the group I introns characterized by Thomas Cech, imagining RNA as catalytic would have been heretic in the context of the RNA knowledge of the 1980s. RNA was thus seen eventually as the scaffolding of the substrate guiding the catalytic action of the protein in a view that corresponded to the ribosome at the time. Indeed, only the enzyme reconstituted from the two separate subunits was active under classical saline conditions (Saenger 1984a) at 10 mM magnesium chloride[4] and 50 mM ammonium chloride. In this context, unraveling the action mechanism of this RNase P thus became a priority for Sidney Altman and his collaborators, especially since in vivo studies undertaken in other laboratories demonstrated at the same time its preeminence in the maturation of many tRNA precursors (McClain 1977).

As with Thomas Cech's discoveries, cloning and in vitro transcription methods enable a better understanding of the enzyme's function. On the one hand, the enzyme's protein is purified from bacterial extracts and, on the other hand, the RNA is produced in vitro by transcription. Obtaining these two molecules makes it possible to test all kinds of incubation conditions for samples containing either the protein or the RNA, or both the RNA and the protein. The results demonstrate the activity of the reconstituted enzyme from the protein and the RNA as well as that the "control" sample containing only the RNA causes the cleavage of the tRNA precursor under highly saline conditions of 60 mM of magnesium chloride and 100 mM ammonium chloride (Guerrier-Takada and Altman 1984, 1986). Refinement of the experiments confirms that RNase P RNA has the characteristics of an enzyme, i.e. it accelerates the reaction and remains unchanged. The same RNA participates in many successive reactions, and it is effective even in small quantities. It is indeed a catalyst since it lowers the activation energy, remaining unchanged during several catalytic cycles.

---

4 The physiological intracellular concentration of Mg ion^{2+} is about 13 mM. It is only 1 mM in blood plasma.

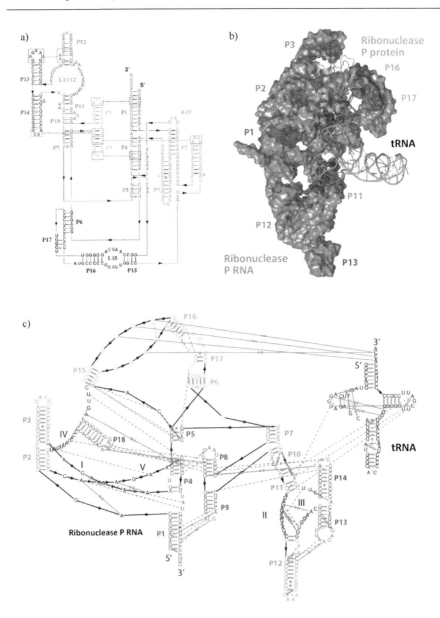

Figure 3.2. *Secondary and tertiary structures of RNase P of* Thermotoga maritima. *For a color version of this figure, see www.iste.co.uk/masquida/ribozymes.zip*

COMMENTARY ON FIGURE 3.2.– a) *The secondary structure as it was deduced from work before the first crystallographic structures on the basis of biochemical data, site-directed mutagenesis and molecular modeling (for review, see Masquida et al. (2010)). This secondary structure is already "borrowed" from 3D in that the helices are placed relative to each other in connection with their relative positions from Christian Massire's molecular model (Massire et al. 1997, 1998). b) The three-dimensional structure of the holoenzyme in the presence of the substrate (Reiter et al. 2010) made it possible to understand the detailed folding of the RNA subunit (blue surface), the interactions between it and the protein (pink) shown in ribbon mode, and interactions with substrate tRNA (green). Regions of the RNA subunit in contact with tRNA are colored red. c) Hydrogen bonds are indicated using the nomenclature of Leontis and Westhof on an expanded secondary structure deduced from the three-dimensional structure. The tRNA (gray) adopts the classical cloverleaf secondary structure. Note that the tRNA does not have a classical anticodon loop but modified by an 11 nt tetra-loop receptor (see Chapter 1) so as to promote crystal contacts. This structure allows for observing the degree of compaction and surface complementarity between the three molecular partners.*

It was not until the efforts of various research groups that the structure of this enzyme was finally available. First, at the end of the 1990s, the structure of the *Bacillus subtilis* homologue of the *E. coli* C5 protein was published (Stams et al. 1998). This small protein of only 12 kDa interacts with the cleavage end of the tRNA precursor and lowers the affinity of the enzyme for the reaction product, thus promoting the release of the tRNA with the mature 5' end (Crary et al. 1998; Niranjanakumari et al. 1998).

The structure of a fragment of the RNA subunit of about 200 nt capable of autonomous folding and conferring the recognition specificity of tRNA precursors, the specificity (S) domain, was solved in the early 2000s (Krasilnikov et al. 2003, 2004), quickly followed by the structure of the entire RNA subunit (Torres-Larios et al. 2005, 2006).

Finally, in 2010, the structure of the holoenzyme with its substrate was solved, revealing the specific interactions between the protein and nucleic subunits and the contacts that each one establishes with the substrate

(Figure 3.2), shedding light on decades of biochemical results that remained difficult to interpret (Reiter et al. 2010).

3.3. The first consequences of these discoveries

In the experiments concerning group I introns and RNase P, the study of RNAs synthesized or isolated in vitro broke the deadlock. Each time, it was the observation of the unexpected behavior of the negative control experiment that overturned the initial conception of the researchers by giving a positive result. At that time, it seemed indeed "heretical" to imagine that nucleic acids could act as enzymatic catalysts like proteins. The results obtained were therefore in total opposition with the knowledge of the time. It was then necessary to these researchers to have a good dose of courage and confidence in their experimental results to present their results to their peers. This quasi-prophetic position when it came to overturning the state of knowledge, to changing the paradigm, was the result of a large number of experimental repetitions as well as the use of different approaches that converged towards the same conclusion for which the author himself was not prepared. It was the author's intimate conviction that pushed him to throw himself into the arena at the risk of being badly treated. Moreover, the first versions of their articles were welcomed with faint-heartedness by the experts of the time. Nevertheless, once published, these two papers widened the field of possibilities, authorizing other scientists to consider that the RNAs they were studying could be endowed with catalytic activity. The world of ribozymes was then within reach.

Although the RNAs of group I and II introns and RNase P are diverse in size, structure and function, the reaction mechanism of catalysis is very similar (Figure 2.4). In the case of group I introns and RNase P, it is the hydroxyl group of an external cofactor that is involved in the initiation of the reaction, a guanosine (3' hydroxyl group) and a water molecule (H-OH), respectively. In the case of group II introns, it is an intron adenosine ("internal" cofactor) whose 2' hydroxyl group plays the same role as the 3' hydroxyl group of guanosine in group I introns. Group I and II introns define the category of self-splicing ribozymes.

3.3.1. *The ribosome, a long-ignored ribozyme*

In its original form, the central "dogma" of molecular biology places RNAs as messengers (mRNAs) between DNA and proteins. When decoding mRNA information through the genetic code, ribosomal RNAs (rRNAs) associate with proteins to form the ribosome (Figure 3.3). When the translation mechanism of messenger RNAs is highlighted, the RNAs of the two ribosomal subunits of several thousand nucleotides in total are perceived as simple scaffolds structuring the 50 or so ribosomal proteins that orchestrate the chemical reactions of amino acid polymerization by reading the code carried by the mRNA. In a simplified view of this process, transfer RNAs (tRNAs) constitute the adaptor molecules of the genetic code by carrying a specific amino acid of the three-letter coding unit of mRNAs, the codons. The ribosome deciphers these codons by bringing a tRNA whose anticodon is complementary to this one on the small subunit (composed of the 16S RNA and about 20 proteins in prokaryotes). The amino acid carried by each tRNA is covalently linked to the previous one by the transpeptidylation reaction that takes place on the large subunit (composed of 23S and 5S RNAs and about 30 proteins in prokaryotes). The iteration of this process results in the synthesis of proteins in the form of linear chains. This process therefore involves RNA "messages" of the genetic code carried by the DNA, tRNA adapters which, by binding to ribosomes, allow the polymerization of amino acids, and ribosomal RNAs whose structural and dynamic characteristics make translation possible.

In the midst of the initial discoveries about ribozymes, structural and biochemical studies of the ribosome, this fascinatingly complex molecular machine, have shown that the ribosome is in fact a ribozyme (Nissen et al. 2000). Fifty years of research were necessary to move from a "proteocentric" to a "nucleocentric" view of translation. This paradigm shift would not have been possible without the research that led to the understanding that the ribozyme activity of RNAs had been maintained to respond to essential cellular functions. This research, a priori unrelated to ribosomes, has therefore been crucial for understanding protein translation.

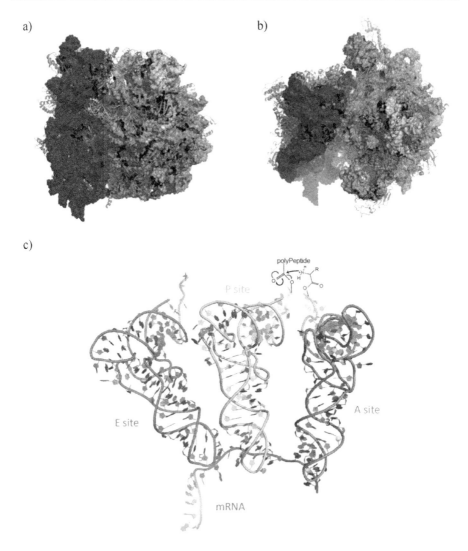

Figure 3.3. *Structure of the elongating ribosome of the thermophilic bacterium Thermus thermophilus (Rozov et al. 2019). For a color version of this figure, see www.iste.co.uk/masquida/ribozymes.zip*

COMMENTARY ON FIGURE 3.3.– *The 16S (blue) and 23S (orange) ribosomal RNAs constitute, respectively, the 30S and 50S subunits with their procession of 56 specific proteins (21 for the 30S subunit and 35 for the 50S). The 5S*

RNA in light green constitutes an important functional region of the 50S subunit. a) Side view of the bacterial ribosome. b) Top view. c) The tRNAs occupying the A, P and E sites (amino-acyl, peptidyl and exit) are at the interface of the subunits. The 3' ends of the tRNAs occupying the A and P sites point to the center (PTC) where the transpeptidylation reaction occurs, the first electron transfer step of which is represented by black arrows and the second step by red arrows in a mechanism in which only 23S RNA is involved (Nissen et al. 2000). Following the nucleophile attack by the amine group, a tetravalent transition state is formed that evolves into the disruption of the ester bond connecting the peptide being synthesized to the tRNA at the P site. After this reaction, translocation of the ribosome to the next codon of the mRNA switches the tRNAs from the A site to the P site, and from the P site to the E site. The tRNA at the E site is released for recycling.

Structural data obtained since the 2000s have revealed the precise location of the region of the ribosome involved in catalysis (Ban et al. 2000; Wimberly et al. 2000; Yusupov et al. 2001). This is the peptidyl-transferase center (PTC), which is located in the domain V of the 23S RNA of the large subunit. This region of the ribosome is composed only of RNA. It is also the target of antibiotic inhibitors of bacterial protein synthesis. The reaction mechanism proposed from structural data obtained in 2000 (Nissen et al. 2000; Polacek et al. 2001; Polacek and Mankin 2005) first suggested an extension of the chemical repertoire of nucleic bases and their reactivity through the formation of tautomeric (Figure 3.4) or alternative protonated forms (Singh et al. 2015), which is indeed the case for most ribozymes. Nevertheless, in the case of the ribosome, it has been proposed that the energy stored in the ester bond between the tRNA and the peptide being synthesized is sufficient for transpeptidylation, as long as the PTC allows the substrates to be oriented through very precise sets of interactions between nucleotides of the ribosome and tRNAs so that the reaction occurs spontaneously. The transpeptidylation reaction is thus accelerated by a factor of 10^5-10^7 (Polacek and Mankin 2005) compared to the spontaneous reaction.

50 Looking at Ribozymes

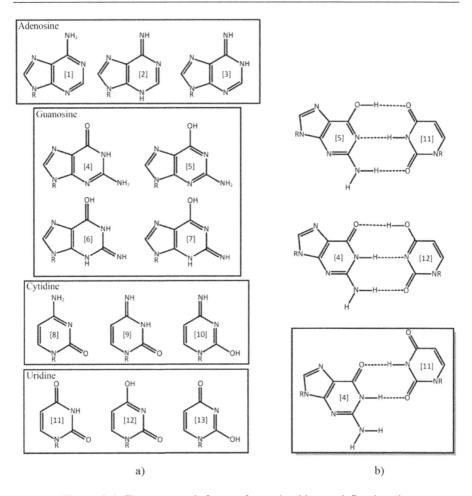

Figure 3.4. *The tautomeric forms of a nucleotide are defined as the different possible arrangements of hydrogen atoms following the rearrangement of the C=C, C=O and C=N double bonds*

COMMENTARY ON FIGURE 3.4.– a) *The four main nucleotides of RNA admit tautomeric forms. R indicates the position of the ribose. [1] 6-amino-A; [2] 6-imino-N3H-A (cis or trans imino function); [3] 6-imino-N1H-A (cis or trans imino function); [4] 6-keto-2-amino-G; [5] 6-enol-2-amino-G; [6] 6-keto-2-imino-G; [7] 6-enol-2-imino-G; [8] 4-amino-2-keto-C; [9] 4-imino-2-keto-C (cis or trans imino function); [10] 4-imino-2-enol-C (cis or trans imino function); [11] (adapted from Singh et al. (2015)). b) Two tautomeric G = U pairings ([5]-[11] and [4]-[12]) adopt canonical, not wobbled, base pair geometries like [4]-[11] typically found in crystallographic structures (shaded framework).*

Tautomeric forms and protonated forms are involved, for example, in the decoding of amino acid-bearing tRNAs by the ribosome. Work by Marat Yusupov demonstrates that the geometry of GoU pairings can be forced to form three hydrogen bonds as in G = C base pairs (Demeshkina et al. 2012; Rozov et al. 2016a). These pairings require the presence of a tautomeric form of guanosine or uridine. Specifically, one of the bases must adopt an enole form to interact with the classical form of the other nucleotide (Figure 3.4(b)). Remarkably, the ribosome architecture strongly constrains the position of the messenger RNA and provides it with an environment that imposes the Watson–Crick *cis* geometry of the first and second base pairs of the codon–anticodon interaction. In the elongation phase, the position of the messenger in the decoding site does not accommodate wobble base pairs in which uridine is shifted to the deep groove side and guanine to the shallow groove side (see Chapter 1 on RNA structure). The stress energy helps to promote the appearance of an enole form, which although rare (ratio of 10^{-4}–10^{-5} to the carbonyl form), allows for a Watson–Crick *cis* three hydrogen bond reaction between tRNA and messenger RNA (Rozov et al. 2018). In the case where an enole form is favored, the ribosome will not distinguish between a cognate tRNA and a pseudo-cognate tRNA because it discriminates them by the morphology of the helix formed between the messenger RNA codon and the tRNA anticodon, which must adopt Watson–Crick *cis* base pairs (Figure 1.11). As long as such a geometry is recognized by the ribosome, the tRNA will be validated and the amino acid it carries will be integrated into the polypeptide chain being synthesized. The ribosome is thus a molecular machine that naturally generates errors in a vital compromise between the speed of synthesis and precision of the mechanism.

These translation errors result in the synthesis of proteins with certain amino acids that are erroneous even though the messenger RNA does not contain mutations. These rare errors do not affect the protein pool and are not genetically heritable. They are unlikely to affect the cell because they are compensated for by the polymerization rate of about 15 amino acids per second. On the other hand, replication or transcription errors due to enole base pairs can in principle affect the integrity of proteins and even generate heritable errors.

Concerning the active sites of ribozymes, the variations of the tautomeric forms of nucleotides participating in the architecture of the catalytic site facilitate the transfer of protons necessary for catalysis. These forms are stabilized by a network of hydrogen bonds and specific weak interactions (stacking, dipole–dipole) and allow the activation of nucleotides that from then become reactive. Tautomeric forms thus extend the repertoire of chemical reactivity of nucleic acids. This idea was already present in the reference article by James Watson and Francis Crick (1953) and taken up in Saenger's book (*Principles of Nucleic Acid Structure*, Saenger 1984b) where tautomers and various protonated forms of nucleotides likely to be involved in DNA mutations (transitions and tranversions) by favoring non-canonical base pairs are described. In 1999, a study conducted on the ribozyme of hepatitis delta virus (HDV) led to the involvement of a protonated nucleobase in the catalytic reaction mechanism (Perrotta et al. 1999). Despite this study, the role of nucleobases in catalysis was not accepted until the mid-2000s with the elucidation of related mechanisms in other families of nucleolytic ribozymes of much smaller size than group I or II introns or RNase P.

3.3.2. *The modified bases*

However, the chemical repertoire of nucleic acids is not restricted to the five nucleotides A, C, G, T and U mentioned so far. There are many nucleotides that show modifications at various positions. These modifications are generally observed on the chemical groups of the nucleobases as well as on the O2' position of the ribose. These modifications can simply see the replacement of an intra-cyclic imino position (N7, for example) by a carbon atom (deaza), or the methylation of a position. But complex modifications such as amino-acylations are also frequently observed. Chemical modifications impact the electrostatic profile of nucleotides, on the one hand, and introduce steric groups that guide very specific recognition mechanisms.

These modifications (Figure 3.5) confer particular functions to RNAs, the most frequently modified being the translational RNAs, tRNAs and ribosomal RNAs. Methylation of an N1 position of adenine, for example, will prevent it from forming a Watson–Crick base pair and thus cannot be involved in a helix. The folding of a tRNA will thus be guided to avoid the formation of alternative three-dimensional structures that might not be recognized by the ribosome. Consequently, the modifications increase the specificity of a tRNA ligand for the synthetase responsible for its correct amino-acylation. Similarly, position 37 of the tRNA anticodon loop is hypermodified (i.e. by a group that generates an important steric hindrance) in order to ensure that the reading of the messenger RNA is restricted to the triplets of nucleotides forming the codons and cannot extend to a fourth nucleotide that would introduce a shift in the reading frame.

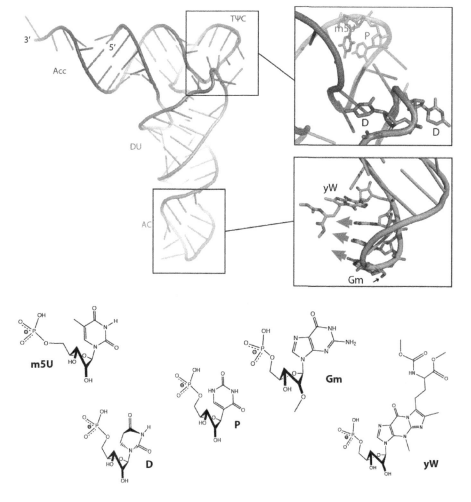

Figure 3.5. *Some modified tRNA bases. For a color version of this figure, see www.iste.co.uk/masquida/ribozymes.zip*

COMMENTARY ON FIGURE 3.5.– *Top left is the three-dimensional structure of yeast phenylalanine tRNA (pdb 1tn1 (Brown et al. 1985)) on which the different domains are indicated (Acc: acceptor arm; DU: dihydrouridine hairpin; AC: anti codon hairpin; TΨC: thymine hairpin). The structures of some of the modified nucleotides present in this tRNA are shown in the insets on the right (oxygen atoms are in red, nitrogens in blue and the color of the carbons vary according to the color of the tRNA domain). Their structural formulas are shown at the bottom of the figure for clarity. At the top, we see the thymine and pseudo-uridine of the TΨC loop in purple. Thymine is nothing more than a methylated uridine at position 5 (m5U). The pseudo-uridine (P) is more surprising because it is no longer the N1 atom of the hexacycle that connects the nucleobase to the ribose but the C5 which has the effect of producing a Hoogsteen face capable of forming two hydrogen bonds. Still in the same inset, in blue, the dihydrouridines (D) of the loop of the same name are represented. The reduction and thus the saturation of the double bond between C4 and C5 of uridine result in the loss of the ring planarity of these nucleobases. Their engagement in stacking interactions is therefore less favored for the benefit of loop formation. The bottom inset shows two modified anticodon loop bases, 2'-O-methyl guanosine (Gm) and wybutosine (yW). In Gm, the 2'-O position is methylated which increases the stability of the loop. The yW is a guanosine hypermodified by an imidazole group on the WC positions onto which is additionally grafted a butanoic acid ester (methyl 4-[3-[(2R,3R,4S,5R)-3, 4-dihydroxy-5-(hydroxymethyl)oxolan-2-yl]-4,6-dimethyl-9-oxoimidazo[1,2-a] purin-7-yl]-2-(methoxycarbonylamino) butanoate). This hypermodification prevents the yW from forming a fourth WC interaction with a messenger RNA nucleotide which would induce a decipherment of the reading frame when deciphering the triplets of nucleotides (symbolized by the gray arrows) that form the codons during translation by the ribosome.*

These modifications have been indexed in databases, including MODOMICS (Dunin-Horkawicz et al. 2006; Boccaletto et al. 2022). Several hundred chemical modifications are listed, and NGS methods have been able to identify them in many types of RNA, including messenger RNAs. We mention this aspect here to indicate that these possibilities open up new fields of research because our knowledge of the functions of chemical modifications in RNAs is still very patchy. We have chosen to simply mention here the question of chemical modifications although this field of research alone would deserve a dedicated book.

3.4. The spliceosome, another ribozyme

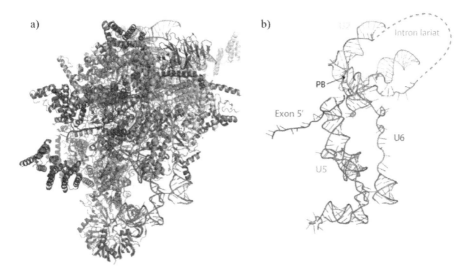

Figure 3.6. *The spliceosome is a molecular machine of extremely complex dynamics (pdb: 5mps (Fica et al. 2017)). For a color version of this figure, see www.iste.co.uk/masquida/ribozymes.zip*

COMMENTARY ON FIGURE 3.6.– *Here, the complex shown corresponds to the 5' exon post-cleavage step. a) The spliceosomal RNAs and the pre-messenger RNA are masked by the 25 protein factors that make up this particular complex. There are about 40 of them in total involved in performing a complete splicing cycle. b) Once cleared of proteins without the change of orientation, the RNAs appear. The branch point of the lariat (BP) is visible near the end of exon 5' (navy blue). The intron-lariat fragments (cyan) visible in the structure are connected by a hatched loop that indicates the path of the complete intron. The 5' exon interacts with loop 1 of the U5 spliceosomal RNA (U5 snRNA). U6 snRNA (orange) stiffens the 5' end of the intron (cyan), while U2 snRNA (yellow) does the same for the part of the sequence that carries the branching nucleotide, adenosine 70. U2 and U6 are intimately linked in this process. Although catalysis relies on*

RNA–RNA interactions, it should be noted that, mechanistically, proteins are needed to structure the complex and ensure its dynamics as well as regulate its activity.

Another macromolecular form of machinery, the spliceosome, has recently joined the ribozyme family after being studied for decades (Figure 3.6). As with the ribosome, the scientific culture built on 1970s paradigms that conferred catalytic properties on proteins alone induced the idea that nucleoprotein machineries were inherently enzymes (Galej et al. 2018). Recent structural studies (Zhou et al. 2014; Hang et al. 2015; Yan et al. 2015; Zhang et al. 2017) of cryo-microscopy show that, as with the ribosome, the active site is composed of RNA and that the proteins are far enough away from it to conclude that they are not involved in transesterification reactions. Still, the proteins are essential for allowing the catalytic RNA part to approach the substrate RNA part corresponding to the intron to be excised from the pre-messenger RNA (Hang et al. 2015), as well as for the regulation of catalysis and its turnover. Excision of the intron requires two transesterification reactions similar to those for group II introns (Costa et al. 2016). In a first step, the adenosine residue internal to the intron is used as a nucleophile to generate the free 5' exon separated from the intron-lariat still connected to the 3' exon. In a second step, the free 3' end of the 5' exon is used as a nucleophile to ligate the 5' and 3' exons. Two metal cations bound to the active site of the RNA play an essential role in catalysis.

3.4.1. *Nucleolytic ribozymes*

3.4.1.1. *The first ribozymes from viruses, virusoids or satellites*

In the 1980s, ribozymes such as self-splicing introns and RNase P were considered relics of a bygone RNA world because metabolic enzymes are proteins. The ubiquity of RNase P and the presence of group I (Haugen et al. 2005a) and II (Bonen and Vogel 2001; Bonen 2012) introns in many organisms, including the mitochondria of plants and fungi, nevertheless posed the question of why these atypical catalysts are maintained in the living world. This question extended to the spliceosome, the molecular machine for splicing in general, consisting of five relatively short ($<$ 100 nt) RNAs (snRNAs) and numerous proteins including Brr2 and Prp8. Did the activity of the spliceosome reside in its RNA or in its proteins? It was not until the mid-1980s that several nucleolytic ribozymes of viroids, virusoids and virus satellites were discovered. These ribozymes also perform a transesterification reaction but using the 2' hydroxyl group as a nucleophile that must be activated beforehand by proton abstraction. The products of the reaction are therefore different from those obtained with introns or RNase P since the nucleophile does not attack the phosphate group from the same angle as the latter. The cleavage reaction products of this S_N2 reaction performed in a different orientation are 5' hydroxyl and 2'–3' cyclic phosphodiester ends (Figure 3.7). In addition to performing the same chemical reaction, these ribozymes additionally share the manner in which they are identified, and also their membership in the category of viral RNAs, viroids, virusoids[5] or virus satellites.

The study of viruses is in fact timeless because they continuously pose problems to public health, agricultural productivity, and zoonoses and new viruses appear very regularly in an unpredictable way. At the beginning of the 1980s, experimental methods were relatively adapted to their study. Genetic material and viral proteins could be extracted in quantity from infected plant or animal tissue. Before the advent of gene amplification techniques (Polymerase Chain Reaction (PCR), for which Kary Mullis received the Nobel Prize in Chemistry in 1993), access to genetic material was directly dependent on the criterion of abundance. Fluorescence microscopy already made it possible to label tissue sections with antibodies. It was in this context that researchers noted that some viruses were accompanied by a viroid or satellite, i.e. a secondary particle that modulates the infectivity of the virus itself or affects the etiology of the disease (Kew et al. 1984). These satellites need the virus to provide the proteins that allow their RNA to replicate. A polymerase runs along the circular RNA ("rolling circle mechanism"; see Figure 3.8) and recopies this template many times, which is composed of circular strands of a few hundred to a few thousand nucleotides. The original strand, called the [+] strand, is copied several times to give a linear strand in which the complementary sequence of the circular strand is repeated. These are called concatamers. The repeats are "cut" and then "ligated" generating circular RNAs of opposite polarity to the original strand by way of a ribozyme present in the sequence (Symons 1989; Flores et al. 2009). The circular complementary [-] strands in turn serve as a template for the polymerase to generate circular "+ strands".

---

5 Virusoids and viroids are circular RNAs that depend on a virus for replication and encapsidation. Viroids do not code for any protein. Satellites are parasites of plant viruses that can be in the form of simple circular RNAs.

58 Looking at Ribozymes

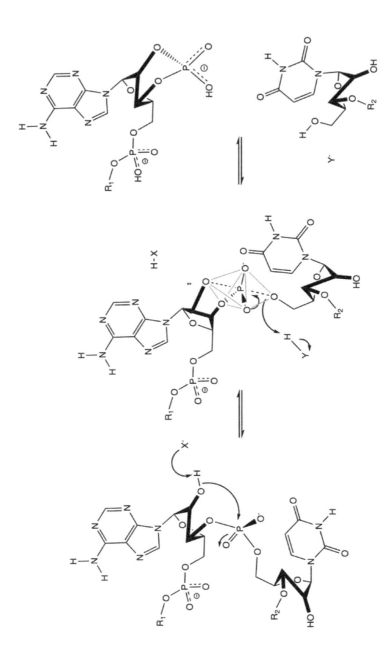

Figure 3.7. *Catalytic mechanism of autonucleolytic ribozymes. For a color version of this figure, see www.iste.co.uk/masquida/ribozymes.zip*

COMMENTARY ON FIGURE 3.7.– *The nucleophilic 2'-OH group is first activated by proton abstraction by a base (X). The electron doublet then attacks the targeted phosphate group following an S_N2 mechanism forming an oxyphosphorane transition state (‡) with a trigonal bipyramid geometry (pink). For the S_N2 mechanism to occur, the three atoms O2', P and O5' must be aligned. The residues between which the cleavage takes place must therefore be distant from each other in order to allow the ribose-phosphate backbone to adopt a favorable conformation. The hydrolysis of the transition state is promoted by an acid that donates its proton to form the 5'-OH group. The 3' end retains the phosphate group which cyclizes onto the 2' and 3' groups (2'–3' cyclic phosphodiester). The production of these products is facilitated by the varied architectures of autonucleolytic ribozymes. This mechanism is related to that of RNase A.*

The "rolling circle" mechanism is deduced from the study of RNA profiles that show the presence of an RNA chain of precise size that occurs in multiples of 2, 3 and up to 6 units, indicating that a long transcript is produced and then segmented into fragments of the same length without the intervention of proteins. Maturation can therefore be obtained in vitro without the help of protein extracts, as was unexpectedly positive in the experiments by Thomas Cech and Sidney Altman. It is therefore ribozymes that are responsible for this maturation.

Viroid ribozymes are characterized as genomic RNA subfragments of plant or mammalian satellites or viroids. The hammerhead ribozyme (Hutchins et al. 1986) corresponds to a region of an avocado sunblotch viroid that causes white spots on green parts of the plant. The hairpin ribozyme (Buzayan et al. 1986) belongs to the RNA of a tobacco viroid that also causes leaf spots (Tobacco Ringspot Virus). Hepatitis delta satellite ribozyme (HδV) (Kuo et al. 1988) is always associated with hepatitis B virus and exacerbates liver disease by producing a specific antigen whose production depends on editing a nucleotide to inosine (a guanosine-derived nucleotide)[6] whose effect is to produce a longer protein that increases the

---

6 Inosine (I) is a modified base comparable to a guanine from which the amino group at position 2 has been removed. This modification only moderately disrupts the secondary structure. I is produced by an adenosine deaminase specific to double-stranded RNA (ADAR: Adenosine deaminase RNA-specific). This enzyme participates in editing, which is an epigenetic modification of RNAs that enables, for example, the modification of C residues into U.

inflammatory response (Chen et al. 2010b). This longer antigen interacts with hepatitis B virus capsid proteins, and allows for the production of hepatitis delta virus particles (Sureau and Negro 2016). Finally, another ribozyme is found in the mitochondria of a natural isolate of a fungus *Neurospora crassa*, the Varkud (VS) isolate (Saville and Collins 1990). These *cis*-cutting ribozymes (i.e. self-cutting) are therefore active only once. The reverse reaction catalyzes the ligation of two ends of different origins (Figure 3.8). In this sense, they are not enzymes in the strict sense, since they do not function iteratively. However, without them, the cleavage/ligation reactions would not occur or would occur too slowly to be involved in any biological mechanism. They therefore significantly accelerated a transesterification reaction at a specific position in the RNA chain.

The characterizations between 1986 and 1990 of these four ribozymes are the first direct consequences of the publication of the work of Cech and Altman.

Because satellite or viroid ribozymes are much smaller than autocatalytic introns or RNase P, their study is facilitated and they offer the opportunity to obtain the first non-tRNA crystallographic structures of RNAs. Although tRNAs present complex structures, they are all globally superimposable, and therefore largely redundant, even though the variations in their structural details linked to differences in sequences and chemical modifications are rich in teachings. The tRNAs must be sufficiently distinct from each other so that the enzyme adding the specific amino acid (the aminoacyl-tRNA synthetases) does not make a mistake, but they must also be sufficiently similar to be recognized by the ribosome that will synthesize the proteins. These features are grouped into identity determinants (which allow for ribosomes to take them on) and identity anti-determinants (which discriminate tRNAs from different tRNA synthetases) (Giege et al. 2012). The first crystallographic structures of ribozymes thus give access within a few years to an expanded view of the structural complexity of RNAs and improve our understanding of folding and catalytic (hammerhead) mechanisms (Pley et al. 1994b; Scott et al. 1995b); (hairpin) (Rupert and Ferré-D'Amaré 2001; Rupert et al. 2002); (HDV) (Ferré-D'Amaré et al. 1998; Ferré-D'Amaré and Doudna 2000) (Figures 3.9A and 3.9B).

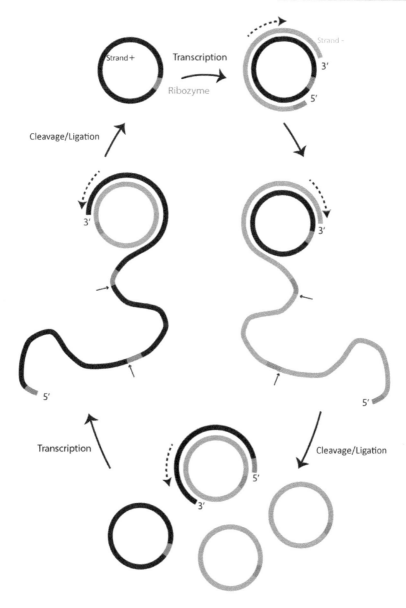

Figure 3.8. *The symmetrical rolling circle replication mechanism. For a color version of this figure, see www.iste.co.uk/masquida/ribozymes.zip*

COMMENTARY ON FIGURE 3.8.– *Genomic RNAs of viroids and virusoids (i) replicate using a rolling circle mechanism in which an RNA polymerase runs along the circular RNA genome (with the + strand in black designated by convention as the most abundant) and retranscribes it many times as many copies of the strand – linked together as concatamers. The RNA folds at the junction of two consecutive genomes as a ribozyme (green region) that cleaves the individual genomes which are then ligated into circular form. Both the original RNA strand (+ strand) and the neo-synthesized strands (- strand) are amplified by a DNA-dependent RNA polymerase. These viral RNAs generally do not encode proteins but both carry a ribozyme which is then referred to as "genomic" or "anti-genomic" depending on which strand they belong to (Jimenez et al. 2015). The model shown in this figure is "symmetric" because the [+] and [-] strands are polymerized and cleaved in the same way. There are other mechanisms where only the neo-synthesized [-] strand remains linear although they contain several genome units (asymmetric mechanism). This strand is then transcribed in turn. The generated [+] strand units are matured by an endonuclease that recognizes a specific stem-loop formed at the end of each genome (Flores et al. 2009). By convention, the [+] strand is the most abundant (Flores et al. 2011).*

3.4.1.2. *Technical developments favor the discovery of new ribozymes*

The routine implementation of PCR (Mullis 1990a, 1990b) in laboratories and, since the early 2000s, the advent of next-generation sequencing (NGS) techniques have led to major advances in understanding living organisms not by growing them but simply by analyzing their DNA or RNA. This is the case of the TARA projects that allow us to understand the diversity of the marine living world (Salazar et al. 2019; Claudet et al. 2020). These methods have also made it possible to sequence complete genomes more and more rapidly and always at lower cost. The exponential increase in the number of sequences available in databases has logically been accompanied by the development of increasingly powerful phylogenetic analysis software and the development of statistical sequence models.

The implementation of these methods has led to the identification of new families of autonucleolytic ribozymes (Barrick et al. 2004; Salehi-Ashtiani et al. 2006; Perreault et al. 2011; Roth et al. 2014; Weinberg et al. 2015) in ever-increasing numbers of organisms, such as in primates and thus in humans (Figure 3.10). It is important to note that the functions of these ribozymes are generally unknown. Beyond their catalytic mechanism and their structure, the study of their role in living organisms is topical. An interesting case that shows how justified this remark is concerns the CPEB3 (cytoplasmic polyadenylation element binding protein 3) protein that seems to be involved in memory mechanisms and whose pre-messenger RNA contains a ribozyme homologous to the one found in the hepatitis delta virus genome (Salehi-Ashtiani et al. 2006; Vogler et al. 2009; Chao et al. 2013; Ford et al. 2019; Qu et al. 2020).

Sequencing methods thus pave the way to genomic exploration. It is possible to search for the presence of ribozymes on the basis of sequences and their variability with statistical models such as the Markov chains implemented in the HMMER program (Eddy 2004). These studies can identify some of these ribozymes in genomes other than viral genomes and verify their expression within these new host genomes (Cole and Lupták 2019).

One prototype nucleolytic ribozyme is the hammerhead ribozyme, which is found throughout the living kingdom, from bacteria to archaea, plants to animals, to humans (Perreault et al. 2011). The recently identified twister ribozyme is equally present in bacteria and eukaryotes (Roth et al. 2014).

Data mining methods also reveal new types of structured RNAs such as riboswitches, which are transcriptional or translational regulatory elements first identified in bacteria and which may carry ribozyme activity such as GlmS RNA which is a riboswitch whose ligand participates in the transesterification reaction (Winkler et al. 2004). Genomics research also indicates that the complexity of genomes is correlated with their content of RNAs that are not linked to a protein-coding phase. For example, the human genome contains only ~1.2% of protein-coding sequences, while at least 98% of the genome is transcribed into RNA (Mattick and Makunin 2005). The "nucleocentric" view is thus reinforced by the observation that genomes are not only protein factories. These observations naturally lead to the discovery of other RNA-centered biological mechanisms such as RNA interference (2006 Nobel Prize in Physiology and Medicine to Andrew Fire and Craig Mello) (Fire et al. 1998; Fire 2007), riboswitches (Mandal et al. 2003; Pavlova et al. 2019) and small regulatory RNAs (Gottesman 2005; Mattick and Makunin 2005), or the CRISPR/Cas system (2020 Nobel Prize in Chemistry to Jennifer Doudna and Emmanuelle Charpentier) (Wang et al. 2022). These mechanisms all allow the regulation of gene expression, and while the presence of ribozymes within riboswitches is attested (Lee et al. 2010), other ribozymes remain to be discovered.

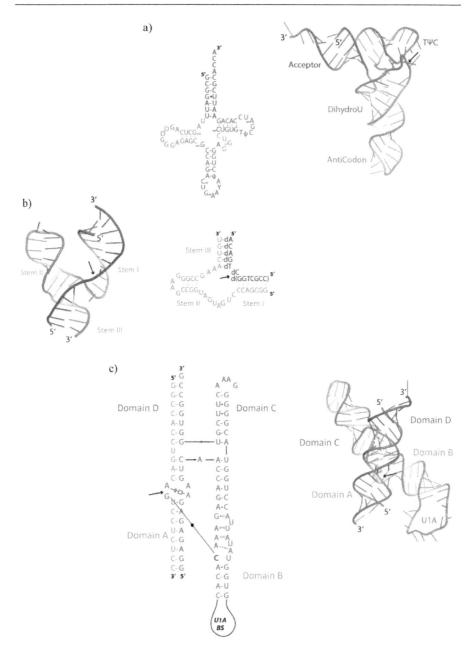

Figure 3.9. *Crystallographic structures of historical ribozymes. For a color version of this figure, see www.iste.co.uk/masquida/ribozymes.zip*

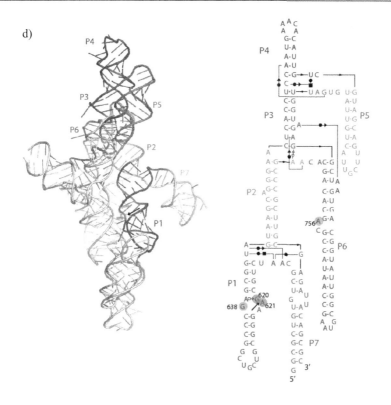
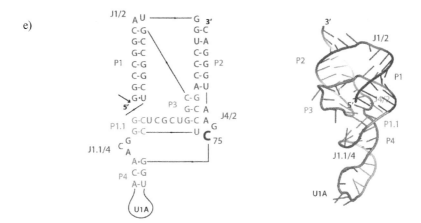

Figure 3.9. *Crystallographic structures of historical ribozymes (continued). For a color version of this figure, see www.iste.co.uk/masquida/ribozymes.zip*

COMMENT ON FIGURE 3.9.– *a) Yeast tRNA^{Phe} (pdb: 1tn1 (Brown et al. 1985)) could be considered a ribozyme because it presents a specific Pb^{2+} atom binding site that catalyzes the cleavage of the ribose-phosphate backbone between residues D17 and G18. The position of the blue strand cleavage is indicated by an arrow as in (b) and (c). There is no arrow in (d) because the HDV ribozyme shown corresponds to the product of the cut that occurs in 5'. The different ribozymes have structures that are quite different from those of tRNAs. In particular, they have more complex inter-helix junctions, loops with varying numbers of residues and pseudoknots. b) The hammerhead ribozyme (pdb:1hmh (Martick and Scott 2006; Martick et al. 2008b)) is organized around a three-helix junction. Its architecture allows for the cleavage of the substrate strand (blue at the base of stem I). c) The hairpin ribozyme (pdb: 1m5k (Rupert and Ferré-D'Amaré 2001)) has an architecture based on a four-helix junction that stacks two by two and forms a characteristic angle of about 60° that promotes the interaction between a guanine of the A domain with the B domain. This constraint gives the backbone a conformation that promotes its rupture. The U1A domain is a particular RNA that interacts with the U1A protein of the spliceosome. This genetically engineered RNA–protein module serves to promote crystallization (Ferré-D'Amaré 2010). d) The ribozyme in the Varkud isolate of Neurospora crassa originates from a mitochondrial satellite plasmid. It was by solving its crystallographic structure that the mystery of its mode of action was solved (pdb: 4r4v (Suslov et al. 2015)). Upon dimerization, this ribozyme cuts the substrate strand of its partner (arrow on the green P1 loop stem). a) and c) are from plant viroids that distinguishes them from the hepatitis delta satellite ribozyme (e) HDV (pdb: 1drz (Ferré-D'Amaré et al. 1998)), which is a virus often associated with hepatitis B in humans.*

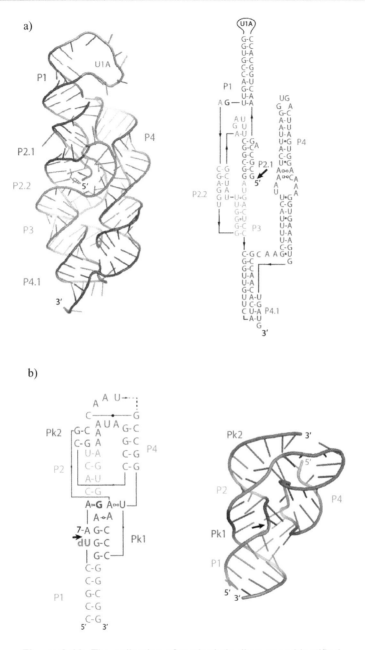

Figure 3.10. *The collection of nucleolytic ribozymes identified since the early 2000s. For a color version of this figure, see www.iste.co.uk/masquida/ribozymes.zip*

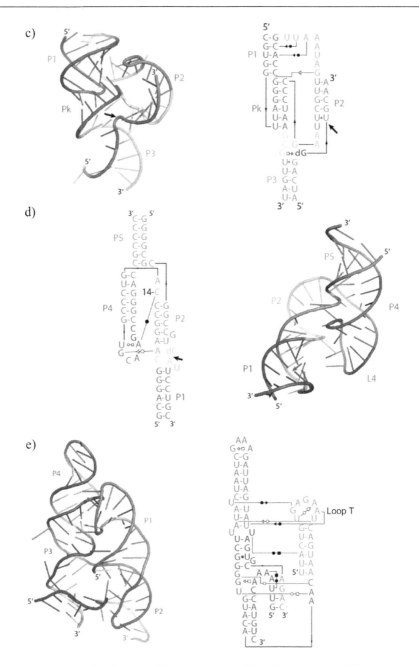

Figure 3.10. *The collection of nucleolytic ribozymes identified since the early 2000s (continued). For a color version of this figure, see www.iste.co.uk/masquida/ribozymes.zip*

COMMENTARY ON FIGURE 3.10.– *a) GlmS, pdb 2h0s (Klein and Ferré-D'Amaré 2006); see Figure 3.9 for explanation of the meaning of the U1A annotation. b) Twister: pdb 4oji (Liu et al. 2014). c) Pistol: pdb 5k7c (Ren et al. 2016). d) Twister sister: pdb 5t5a (Liu et al. 2017). e) Hatchet: pdb 6jq5 (Zheng et al. 2019). Black arrows indicate the region of the ribose-phosphate backbone that is cut. The absence of an arrow for (A, helix P2.1) and (E, helix P1) indicates that the resolved structure corresponds to the cleavage product that occurs in 5'. In the case of GlmS, glucosamine-6-phosphate participates in the reaction through its amine moiety, which donates a proton to promote the formation of the 5'-hydroxyl leaving group (Gong et al. 2011). The metabolite synthesized by the enzyme whose expression is under the control of the ribozyme glmS thus participates in catalysis. Except for GlmS, which regulates the expression of the glutamine-fructose-6-phosphate amidotransferase gene that produces glucosamine-6-phosphate (Winkler et al. 2004), the roles of the other four ribozymes in this figure have not been clearly identified.*

In silico research methods based on NGS sequence mining have led to the discovery of riboswitches and then more recently to genomic ribozymes that had never been characterized as twister, twister sister, hatchet and pistol (Weinberg et al. 2007; Nawrocki and Eddy 2013). The functions of these ribozymes, which constitute new families of autocatalytic ribozymes, are not yet elucidated (Weinberg et al. 2015). Longer known ribozymes, such as the hammerhead ribozyme, are also found in the novel context of circular RNAs with emerging functional implications (Andersen et al. 2016). Inserted into retrotransposon-like circular RNAs, hammerhead ribozymes give rise to a new functional family of RNAs, the retrozymes (Cervera et al. 2016). In this context, these ribozymes also catalyze a reversible transesterification reaction. As in the case of viroid genomic RNA replication, this transesterification involves cutting the RNA chain at a point that may be right in the ribozyme sequence or at one of its 5' or 3' ends. The cleavage thus affects the structural integrity of the ribozyme. The most immediate function of these genomic ribozymes is therefore probably to separate functional domains of RNAs that function in concert to render them inoperative or to induce their degradation. However, it is striking that the function of this transesterification reaction is still unknown for the majority of the nine known nucleolytic ribozyme families. Moreover, their catalytic mechanisms are still debated.

3.4.1.3. *In vitro selection opens a window to the unknown*

The appearance of another experimental method further increases the possibilities of scientific discoveries by allowing Darwinian evolutionary experiments to be performed in vitro. This method imagined by Jack Szostak, Gerald Joyce and Larry Gold (Ellington and Szostak 1990; Robertson and Joyce 1990; Tuerk and Gold 1990), named SELEX for Systematic Evolution of Ligands by EXponential enrichment, enables selecting RNAs for their specific substrate recognition or reactivity properties from a library of molecules obtained by combinatorial chemistry (Figure 3.11 (Ellington and Szostak 1990)). This method has demonstrated that the chemical reactivity of RNAs could be extended in vitro to the repertoire of reactions usually devolved in vivo to protein enzymes alone. The chemical reactivity of RNAs is therefore probably not restricted to transesterification as the ribozymes identified so far in nature might suggest. The discovery of the recognition of all kinds of substrates such as proteins, as well as vitamins or metabolites by RNAs called aptamers, has prepared the discovery of riboswitches. These can be defined as aptamers (*apt* (Latin): capable; *meros* (Greek): part) coupled to an expression platform that undergoes a conformational change depending on the presence of the substrate. This expression platform can be an inhibitor or activator of transcription, translation or splicing, but this list is not exhaustive. As riboswitches are not the subject of this book, we invite the reader to turn to specific articles (Mandal et al. 2003; Winkler et al. 2003; Tucker and Breaker 2005).

The SELEX method applies specific selection pressures to natural ribozymes to generate ribozymes with new properties, so-called "augmented" ribozymes. The activity of these mutant ribozymes will depend, for example, on the temperature or the binding of a ligand. There are many examples of this type of application (Soukup 2006). Fully artificial ribozymes are thus generated with a wide repertoire of catalytic properties. This repertoire ranges from cleavage, hydrolysis, ligation, polymerization and almost all the usual reactions performed in RNA metabolism, to more exotic reactions such as methyl or acyl group transfer, amide bond formation or Diels–Alder reactions (Serganov et al. 2005). Keep in mind that the only natural ribozyme known to have survived through time, and capable of performing a reaction not based on phosphate transesterification, is the ribosome, which catalyzes the formation of the peptide bond from an electrophilic ester and a nucleophilic amine.

The SELEX method thus provides a proof of concept of the RNA world by characterizing artificial ribozymes that perform enzymatic reactions previously reserved for proteins. Ribozymes involved in major RNA metabolism reactions have been selected with polymerase and ligase activities (Shechner et al. 2009; Shechner and Bartel 2011; Attwater et al. 2013), the ribonuclease activity being already present in the natural ribozymes discovered so far. All three activities are fundamental for the autonomous replication of pathogens such as viroids presented as relics of the RNA world. Some current viroids conserve the ribonuclease activity carried by the ribozymes of their genome and antigenome.

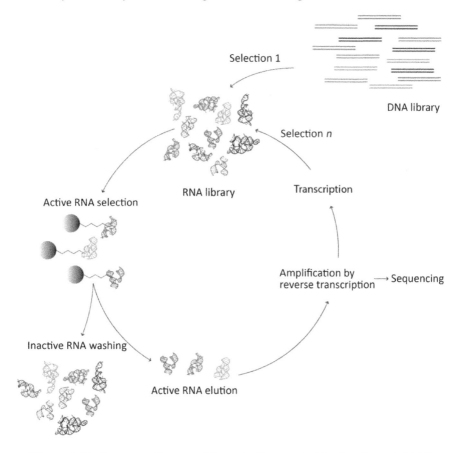

Figure 3.11. *Summary diagram of the selection method in vitro (Luptak 2016). For a color version of this figure, see www.iste.co.uk/masquida/ribozymes.zip*

COMMENTARY ON FIGURE 3.11.– *The initial selection step (selection 1) begins with an RNA population obtained by transcription of a DNA library from which many consecutive positions have been synthesized in solid phase from a mixture of the four nucleotides. If n consecutive residues are synthesized in this way, then there are 4^n different possible sequences. In order for the DNA population to represent the desired diversity, or in other words for each sequence to be represented in the starting library, large mass libraries of the order of a milligram are generally required. The starting RNA population is subjected to a first biological screen which involves fixing the RNAs by affinity for a substrate or following a chemical reaction to a material such as an agarose bead which can be retained by sieving. The non-affine or unreacted RNAs are removed by rinsing. The retained RNAs are then eluted from the experimental device to be sequenced following a reverse transcription step. This new library, whose sequence diversity has largely decreased but which presents a much higher activity, is amplified by a reverse transcription step followed by an amplification by PCR (RT-PCR). After transcription, the resulting RNAs are selected again in the next cycle. In each cycle, parameters such as incubation times, ionic concentrations and pH can be modified to increase the stringency of the conditions and obtain more and more specific RNAs.*

4

Ribozyme Engineering and the RNA World

With the advent of PCR and in vitro selection (SELEX), it has become possible to easily clone all kinds of natural ribozymes and study their catalytic properties based on their nucleotide sequence (Sargueil and Burke 1997; Jabri and Cech 1998; Lau Ferré-D'Amaré 2016). These ribozymes can also serve as therapeutic molecules expressed in vivo by directing their cleavage activity against target RNAs (Scarborough and Gatignol 2015; Zorzan et al. 2015; Felletti and Hartig 2017; Asha et al. 2018; Lee et al. 2018; Moore et al. 2019; Tsiamantas et al. 2019; Duan et al. 2020) to which they hybridize through sequence complementarity. Because of their small size, nucleolytic ribozymes are valuable molecular biology tools for controlling gene expression, recognizing another RNA molecule that becomes the substrate (Uhlenbeck 1987).

SELEX enables characterizing the structure/function relationships of ribozymes and developing artificial ribozymes that catalyze reactions distinct from transesterifications of the phosphate groups of nucleic acids (Talini et al. 2009) (Figure 4.1). The identification in ribozymes of nucleotides essential for catalysis allows this knowledge to be used for the design of new ribozymes (Sargueil et al. 2003). A ribozyme capable of performing the typical cycloadditions of Diels–Alder reactions has been developed (Frauendorf and Jaschke 1998; Keiper et al. 2004; Serganov et al. 2005) or ribozymes (called flexizymes) capable of transesterifying amino acids that modify the genetic code and enable access to powerful synthetic biology methods (Suga et al. 1998; Murakami et al. 2006a, 2006b; Xiao et al. 2008). Originally, nucleolytic ribozymes are both enzyme and substrate. With the knowledge acquired, residues essential for catalytic activity can be conserved within an RNA molecule that acts as the enzyme that can act in *trans* after recognizing an RNA target. One of the most emblematic examples of these technological developments is certainly the flexizyme. This artificial ribozyme evolved by SELEX allows, among other things, the grafting of a natural or artificial amino acid to a selected tRNA provided that it has an aromatic ring with six carbons that will serve as a leaving group during the addition reaction (Murakami et al. 2006a, 2006b; Ohuchi et al. 2006). This property allows the genetic code to be modified and increases the number of artificial amino acids that can be used in an in vitro translation experiment. New non-natural peptides with human health benefits are thus obtained relatively easily. These technological developments have led to the development of a biotechnology company called Peptidream (https://www.peptidream.com), which has sold licenses to most of the major pharmaceutical companies. The company is listed on the stock exchange for several billion dollars.

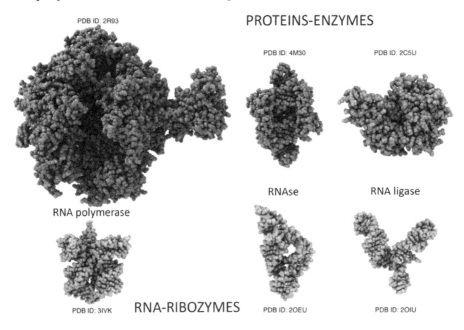

Figure 4.1. *Catalytic activities required for viroid replication carried by host enzymes or ribozymes. Only RNase activity, carried by a ribozyme, is present in some viroids; the other two activities were obtained with artificial ribozymes generated by SELEX (3ivk (Shechner et al. 2009; Piccirilli and Koldobskaya 2011)); (2oiu (Robertson and Scott 2007)). For a color version of this figure, see www.iste.co.uk/masquida/ribozymes.zip*

Diverted from their original function, nucleolytic ribozymes are modified for use as molecular scissors long before microRNAs and silencing mechanisms (Fire et al. 1998; Fire 2007).

Subsequently, they can be combined with other RNAs, with the objective of coupling specific functions to produce new biological mechanisms in a synthetic biological perspective. Combined with RNA motifs or domains with properties such as ligand binding or temperature (Giuliodori et al. 2010) or pH sensitivity (Nechooshtan et al. 2009; Pham et al. 2017), ribozymes will be able to transform into regulatory riboregulators and control a gene as a function of temperature, pH or ligand concentration (Penchovsky 2014).

4.1. Classification of ribozymes

Ribozymes are very diverse with respect to the criteria used previously to characterize them in terms of size, structure, folding, function, origin, evolution and catalytic mechanism. Their use for bioengineering purposes is also very diverse since ribozymes can in principle be adapted to various targets. The first ribozymes discovered, group I introns and RNase P, are large ribozymes (several hundred nucleotides) as opposed to small nucleolytic ribozymes (less than a hundred nucleotides). Group I or II introns are players in RNA maturation by self-splicing, as is the spliceosome derived from group II introns, as are some protein cofactors derived from proteins translated from the coding phases of these introns (Galej et al. 2018). Other ribozymes such as the lasso-capped ribozyme (Meyer et al. 2014) are also involved in a maturation function, that of maturation of the coding messenger for an endonuclease that cleaves the ribosomal pre-RNA of the small subunit at position 956 (Johansen and Vogt 1994). RNase P is involved in the maturation of transfer RNAs by hydrolysis of the 5'-sequences of precursors. Table 4.1 summarizes the different classes of natural ribozymes listed in RFam.

RFam ID (Kalvari et al. 2021)	Family/description	Size (nt)	Initial reference
RF00028	Group I autocatalytic introns	~300	Cech et al. (1981)
CL00102	Group II autocatalytic introns	~400	Michel et al. (1982)
CL00010	Hammerhead ribozyme	~50	Hutchins et al. (1986)
RF00009	Nuclear RNase P	~300	Koski et al. (1976)
RF00010	Bacterial RNase P class A	~400	Guerrier-Takada et al. (1983)
RF00011	Bacterial RNase P class B	~400	Guerrier-Takada et al. (1983)
RF00030	RNase MRP	~300	Chang and Clayton (1989)
RF00373	Archaeal RNase P	~400	Darr et al. (1990)
RF00094	Hepatitis delta virus ribozyme	~80	Kuo et al. (1988)
RF00173	Hairpin ribozyme	~50	Buzayan et al. (1986)
RF00622	Mammalian CPEB3 (HDV-like) ribozyme	~70	Salehi-Ashtiani et al. (2006)
RF01807	GIR1 branching ribozyme or LC ribozyme	~200	Johansen and Vogt (1994)
RF00234	GlmS ribozyme	~80	Winkler et al. (2004)
CL00120	Twister ribozyme	~60	Roth et al. (2014)
RF02678	Hatchet ribozyme	~80	Weinberg et al. (2015)
RF02679	Pistol ribozyme	~70	Weinberg et al. (2015)
RF02681	Twister sister ribozyme	~70	Weinberg et al. (2015)

Table 4.1. *Classification of ribozymes according to RFam (Gardner et al. 2011; Kalvari et al. 2021). RF stands for RNA family and CL stands for RNA Clan. The only ribozyme that is not represented in RFam is the VS ribozyme because only one occurrence has been described to date, not allowing its classification as a family*

4.2. Classification of ribozymes according to catalytic mechanism

Ribozymes such as group I or II "self-splicing" introns function differently from autonucleolytic ribozymes. In addition to their size, this is based on the observation that self-splicing ribozymes have a two-step catalytic transesterification mechanism that depends on metal cations as essential cofactors, whereas the reversible transesterification reaction of autonucleolytic ribozymes instead relies on tautomeric forms of nucleobases that promote the stabilization of reaction intermediates and end products. Nucleolytic ribozymes all catalyze the same transesterification reaction, such as RNase A and related ribonucleases. Nevertheless, their active structures are diverse and thus result from different dynamic folding processes. Each represents an adaptation to a distinct context to operate the same reaction. The glmS ribozyme is unique in that its enzymatic activity depends on its ligand, glucosamine-6-phosphate (Glc-N6P), which acts as a cofactor (Klein and Ferré-D'Amaré 2006; Klein et al. 2007; Gong et al. 2011). This particular property makes this RNA both a ribozyme and a riboswitch (RNAs lacking catalytic activity but whose binding of a key metabolic ligand induces a conformational change in the RNA (Jones and Ferré-D'Amaré 2017)). Among the biological functions in which ribozymes are involved, we can therefore mention the maturation of messenger RNAs via splicing in particular, the maturation of non-coding RNAs, replication in which different viroid ribozymes are involved or the regulation of gene expression via the binding of ligands as cofactors. Other examples include the coupling between riboswitches and spliceosomal introns (Cheah et al. 2007) or group I ribozymes (Lee et al. 2010) that regulate splicing in response to a metabolic signature. Ronald Breaker (Yale University), who identified riboswitches, also showed for the first time their coupling with processes other than transcription or translation (Lee et al. 2010).

In the latter example, an allosteric group I intron contained in a virulence gene of the pathogenic bacterium *Clostridium difficile* has its splicing ability controlled by a cyclic di-GMP riboswitch (di-guanosyl 5' cyclic monophosphate; c-di-GMP). GMP dimers cyclized by two 5',3' linkages accumulate through the action of cellular diguanylate cyclases stimulated by variations in the direct environment of bacteria. The increase in the intracellular concentration of c-di-GMP is detected by riboswitches present in several genes upstream of transcripts encoding adhesins, substrates of enzymes called sortases that allow their attachment to the bacterial cell wall. These adhesins themselves may be involved in anchoring to host tissue (Peltier et al. 2015). This second messenger therefore induces the formation of biofilm or virulence systems (Romling and Amikam 2006). The mechanism by which the c-di-GMP concentration induces alternative splicing by the group I intron of the CD3246 adhesin transcript is as follows. At low concentrations of c-di-GMP, the riboswitch upstream of the group I intron separating the ribosomal binding site (RBS) from the AUG initiator codon produces an aberrant cut that results in the truncation of the messenger, which is then not translated because it cannot be retained on the ribosome. At high concentrations, the situation is reversed. The folding of the riboswitch promotes the correct folding of the intron and produces a two-step splicing of the transcript, which sees the RBS and the initiator codon reunited and the adhesin expressed. This is relocated to the cell wall and allows biofilm formation. The coupling of the riboswitch to this allosteric group I intron induces the gene to be under the control of two metabolites, c-di-GMP, on the one hand, and GTP, on the other hand. The c-di-GMP controls the transition from a motile lifestyle (individual cells) to a sessile lifestyle (aggregated cells).

Finally, the earliest natural nucleolytic ribozymes (ribozyme hammerhead, HDV, hairpin, VS) and those identified more recently (twister, twister-sister, pistol, hatchet) (Weinberg et al. 2010, 2015; Roth et al. 2014) and found in phylogenetically distant organisms (Salehi-Ashtiani et al. 2006; Lunse et al. 2017) are distinct from the artificial ribozymes generated by SELEX (Famulok and Mayer 2014; Hollenstein 2019). The enzymatic variety of these artificial ribozymes (and especially those essential to replication) is an argument for the RNA world as well as a source of inspiration in bioengineering and synthetic biology (Balke et al. 2014). Natural and artificial ribozyme variants have also been generated by SELEX (Berzal-Herranz et al. 1993; Saksmerprome et al. 2004; Ryckelynck et al. 2015). In some cases, RNA fragments with specific reactivity have also been used to create new ribozymes, as in the case of the leadzyme (Figure 3.9(a)), obtained from a lead-reactive transfer RNA fragment (Pan and Uhlenbeck 1992). This ribozyme emphasizes that the presence of the 2'OH group of the ribose provides an opportunity for intrinsic reactivity in RNAs as well as stabilization of conformations favorable to the phosphodiester chain cleavage reaction (Wedekind and McKay 2003).

Ribozymes can be distinguished according to the context of their discovery, the progress of knowledge on the origins of life and the evolution of living organisms (RNA world, RNA–protein world), technological developments (cloning/sequencing, PCR, SELEX) and their use in biodiagnostics and in the fields of health and biotechnology. One ribozyme in particular has attracted constant interest for several decades, the hammerhead ribozyme, of which there are numerous variants both in viroids and in various organisms throughout the living kingdom, from bacteria to eukaryotes, including humans, and including archaea. Because of its size of only ~50 nucleotides, it has also been the subject of numerous enzymological and structural studies. Its reaction mechanism is relatively well understood even though details of the reaction are still actively discussed between specialists, and its study raises issues associated with all ribozymes. Among these recurrent issues, we can mention the origin of catalysis and the dependence on cofactors, the identification of "activated" nucleobases in tautomeric or protonated form, the dynamics of folding and structure, the specific impact of tertiary interactions between distant residues in the sequence, the biological roles possible in a context other than that of viroids and their use in bioengineering. In this respect, hammerhead ribozymes represent a prototype for the study of catalytic RNAs and illustrate the controversies that punctuate studies on these ribozymes.

5

Structures of Ribozymes

The aim of this chapter is to present the structures of some model ribozymes in order to understand their commonalities, their distinctiveness, and how recurrent structural motifs play structuring roles in different architectural contexts. Catalytic reactions are also discussed.

5.1. Structures and catalytic mechanisms of ribozymes

5.1.1. *Hammerhead ribozymes*

The circumstances of the initial discovery of a hammerhead ribozyme (HMH) were outlined in the introduction. Here, we wish to describe its three-dimensional structure by including the architectural motifs that facilitate its catalytic activity. Figure 5.1 shows two structures obtained by radiocrystallography (Scott et al. 1995a, 1995b; Martick and Scott 2006).

The first structure (Figure 5.1(a)) is that of a catalytic minimal construct that was originally characterized in the laboratory of Olke Uhlenbeck in 1990 (Fedor and Uhlenbeck 1990). This construct with the loops closing the h1 and h2 helices removed had the advantage of being formed from two RNA strands, the ribozyme strand (encompassing h3) and the substrate strand. This design facilitated catalytic studies since mutations or chemical modifications could be introduced by chemical synthesis into these relatively short strands. For many years, studies of the catalytic mechanism of hammerhead ribozymes were conducted on this minimal construction.

For a color version of all the figures in this chapter, see www.iste.co.uk/masquida/ribozymes.zip.

Several crystallographic structures of this form were solved under different salt conditions, at different pH values, and with subtle sequence differences, but none provided an unambiguous understanding of the catalytic mechanism (Pley et al. 1994b; Scott et al. 1995b, 1996; Murray et al. 1998b, 2000; Dunham et al. 2003).

Indeed, catalytic studies identified a nucleophilic substitution mechanism passing through a transition state where the phosphorus atom is transiently pentavalent (S_N2, oxyphosphorane transition state in trigonal bipyramid, see Figure 3.7). This mechanism assumes that the three atoms constituting the nucleophile, the electrophilic center and the leaving group, respectively, are aligned. However, none of the crystallographic structures of the minimal form highlighted these characteristics. The question of the nature of the triggers of the reaction thus remained mysterious.

Fortuitously, new examples of hammerhead ribozymes were identified in 2000 (Ferbeyre et al. 2000). Studies of these new forms proposed that the loops establishing tertiary contacts appeared to be responsible for a significant increase in catalytic activity (Canny et al. 2004). The ionic concentration, especially of magnesium ions, could be largely decreased without affecting the catalytic properties (Canny et al. 2004). The crystallographic structure revealed the nature of the contacts between the loops (shown in light blue in the 3D panel of Figure 5.1(b)) (Martick and Scott 2006).

These contacts stabilize a higher energy conformation, i.e. a precatalytic conformation in which the interactions taking place in the triple junction of the minimal form are no longer observed. It is therefore this energetic compensation by the interactions between the loops that activates the triple junction's catalytic site. The conformation of the catalytic site exhibits a near-perfect alignment of the three atoms involved in the nucleophilic substitution, providing at the same time structural evidence to the mechanism of the reaction deduced from the biochemical data (Martick and Scott 2006; Nelson and Uhlenbeck 2006; Anderson et al. 2013; Mir et al. 2015).

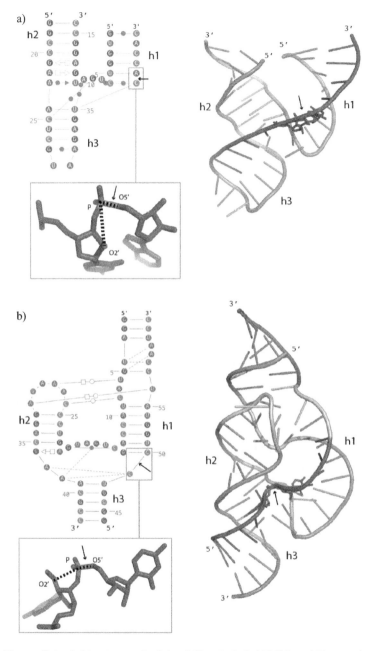

Figure 5.1. *a) Structures of minimal (Scott et al. 1995b) and b) complete (Martick and Scott 2006) forms of hammerhead ribozymes*

COMMENTARY ON FIGURE 5.1.– *The minimal form comes from the sequence of avocado sunblotch virus (ASBV) and that of the complete form of the trematode* Schistosoma mansoni, *which cleaves about 1,000 times faster (1,000 min-1) than the minimal form (Canny et al. 2004). Each panel shows the secondary structure (generated with Assemble software (Jossinet et al. 2010)) where tertiary interactions according to the nomenclature of Leontis and Westhof (2001), the three-dimensional structure and the dinucleotide cleaved during the reaction (inset). The cleaved bond is indicated by an arrow. The relative positions of the three atoms involved in the S_N2 reaction are visualized with hatched lines. The alignment of these, which conditions the reaction, is favored in the structure of the complete ribozyme by the establishment of tertiary interactions between the loops of the h1 and h2 helices. In order to inhibit the in vitro cleavage, a methyl group was added to the nucleophilic O2' group. This modification replaces the O2' proton with a methyl group which prevents the activation of the reaction.*

To better understand the catalytic mechanism of the hammerhead ribozyme, the precatalytic conformation imposed by tertiary interactions between stem-loops I and II of an artificial ribozyme (Saksmerprome et al. 2004) serves as a starting point for structural studies. However, the precatalytic structure has too low an energy to ensure that the relative positions of the ions and nucleobases that may be involved can be observed. To gain access to the structure of a ribozyme transition state analogue, a strategy initially developed on the hairpin ribozyme is to form a complex between the 5' and 3' cleavage products via the intermediary of a vanadate ion (Rupert et al. 2002). The vanadate ion connects the two cut strands by coordination bonds involving the hydroxyl groups (2' and 3', on the one hand, for the upstream cleavage product and 5', on the other hand, for the downstream cleavage product) at the ends of the cleavage products (Figure 5.2). These complexes form a trigonal bipyramid geometry characteristic of the oxyphosphorane transition states of ribozymes. This strategy applied to the hammerhead ribozyme leads to the observation of the transition state's direct environment into sharper focus (Mir and Golden 2016).

Structural analysis and biochemical data allow the deduction that the cleavage reaction results from a mechanism involving a base and an acid. The base, constituted by G12, abstracts the proton from the O2' group, generating the nucleophile that attacks the phosphorus atom of the scissile phosphate group. A hydrated divalent ion serves as a Lewis acid and thus allows a water molecule or the O2' moiety of G8 to generate the 5' hydroxyl group of the leaving group (Figure 5.2).

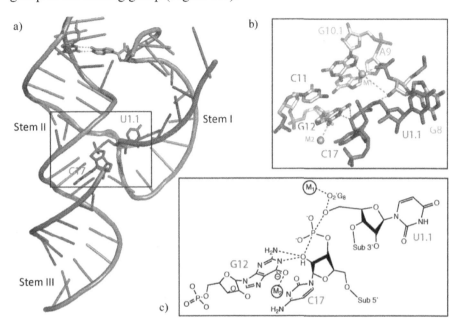

Figure 5.2. *The catalytic mechanism of a hammerhead ribozyme*

COMMENTARY ON FIGURE 5.2.– *a) This "ribbon" representation of a ribozyme (pdb 5eao (Mir and Golden 2016)) shows how the tertiary interaction between stems I and II induces the interruption of the helical continuity of the substrate strand (magenta) that materializes as a "break" between the nucleotides that are bound to the phosphate moiety transesterified by the catalytic action. b) An enlarged view of the active site region shows the interactions involved in the stabilization of the pentavalent transition state represented here by the complex of a vanadium ion with the 2', 3' and 5' hydroxyl groups involved in the catalysis. The magnesium ion M2 promotes the tautomerism of G12, which then behaves as the weak base activating the O2' group of C17. On the other side of the active site, M1 lowers the pKa of the 2' hydroxyl group of G8, which becomes acidic enough to donate its proton to the leaving group, the 5' oxyanion of U1.1 (Wilson et al. 2019). c) The diagram of the active site interpreted from the 3D structure visualizes the transition state in the natural ribozyme. The hatched P–O bonds are either being cleaved (O5'–P) or being formed (O2'–P).*

Research on not only hammerhead ribozyme but also other endonucleolytic ribozymes has identified four important characteristics of ribozyme catalytic reactions. These features were formalized by Scottish researcher David Lilley (Dundee University) and are stated as follows (Lilley 2017): (i) the overall architecture of the ribozyme must facilitate the alignment of the three atoms involved in type 2 nucleophilic substitution (S_N2); (ii) the oxyphosphorane transition state must be stabilized by the establishment of additional hydrogen bonds and the additional negative charge must be compensated for by an ion or mesomeric forms; (iii) the nucleophilic 2' hydroxyl group must be activated by a weak base which often happens to be a neighboring nucleotide; (iv) a weak acid must donate a proton to the 5' hydroxyl leaving group to promote its departure.

In the case of hammerhead ribozyme, these different points are more or less identified, even though doubts remain, especially concerning the role of the ions, especially for the one that seems to be involved in the neutralization of the 5' hydroxyl group of the leaving group (characteristic iv). The most recent work from David Lilley's laboratory confirms the involvement of G12 as a general base, but specifies that it is the hydroxyl 2' of G8 that is the general acid involved in the stabilization of the oxyanion 5' of the leaving group. Nevertheless, the presence of a magnesium ion is required to sufficiently lower the pKa (Wilson et al. 2019). These questions, occurring many years after the publication of the original crystallographic structures, show the difficulty of crystallizing a ribozyme in its active form while modifying certain chemical groups to inactivate them and give them time to crystallize over several days. Moreover, identifying the ions with certainty in crystallography is very difficult and replacing them with other elements of the periodic table does not guarantee their functional equivalence, which can lead to overinterpretations (Auffinger et al. 2020). Therefore, extreme caution should be exercised when describing the catalytic mechanisms of ribozymes. Studies of the catalytic mechanisms of different ribozymes have identified the nucleotides involved in the reaction as base or acid. We present below the examples corresponding to two classic ribozymes, the hairpin ribozyme and the glmS riboswitch.

5.1.2. *The example of the hairpin ribozyme*

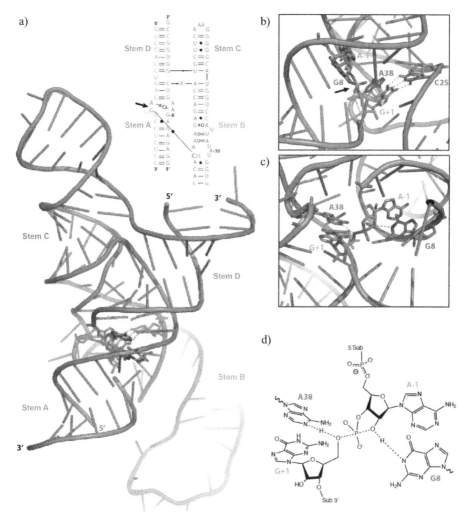

Figure 5.3. *The structure of the hairpin ribozyme and its catalytic mechanism*

COMMENTARY ON FIGURE 5.3.– *a) The overall structure of the hairpin ribozyme is in the form of four helical domains stacked two by two to form an "X". The secondary structure visualizes how the substrate strand (orange) is paired with the ribozyme strand (blue) and forms the inner loop of the A domain. The C25 interaction that allows the extraction of G_{+1} from the nucleotide stack of the A-loop materialized using the nomenclature of*

Leontis and Westhof (2001) is also shown in 3D. b) The arrow points to the bond cleaved during the reaction. c) The transition state structure mimicked by the vanadate complex (pdb 1m5o (Rupert et al. 2002)) is oriented to clearly visualize how the nucleotides of the substrate strand A_{-1} and G_{+1} are splayed apart in order to release the vanadium complex and thus allow G8 and A38 to play their respective roles as base and acid. d) The theoretical scheme in which the vanadium complex is replaced by the oxyphosphorane involved in the catalytic reaction materializes the bonds being cleaved or formed by dotted lines.

The hairpin ribozyme (Buzayan et al. 1986) appears in vivo as four helices that stack two by two to form a chiasma (McKinney et al. 2004), reminiscent of the Holiday junction seen in DNA during genetic recombination events (Lilley and Norman 1999) (Figure 5.3(a)). The A and B stems contain internal loops that interact in a manner that promotes catalysis. As with the hammerhead ribozyme, a more flexible form was initially used as a model ribozyme to study catalytic parameters and the reaction mechanism (Murchie et al. 1998). This shape, circumscribed to the A and B stems, although minimal in size, had a disadvantage in crystallization, a domain in biology where having molecules with stable structure is required. The chiasma shape has elucidated the crystallographic structures of several reaction intermediates and thus shed light on many issues related to catalysis (Rupert and Ferré-D'Amaré 2001; Rupert et al. 2002). In particular, the pairing of C25 from the loop in loop B with the G_{+1} nucleotide that undergoes cleavage in loop A causes the alignment of the three implicated atoms in S_N2 (Figure 5.3(b)).

The transition state oxyphosphorane is stabilized by additional hydrogen bonds and the nucleotides involved in the acid/base mechanism of nucleophile activation and leaving group stabilization are unambiguously identified (Figure 5.3(c)). These are the N1 imines of G8, on the one hand, which abstracts the proton from the O2' group of A_1, and A38, on the other hand, which stabilizes the O5' leaving group of G_{+1}. Note that the mesomeric states of G8 and A38 adopt their most stable form only after they have fulfilled their function. The starting mesomeric forms are very unstable and must therefore be promoted by a particular structural context involving ions and/or local pKa changes. These forms are difficult to observe crystallographically. The protons are directly observable only at resolutions >1 Å, and the mesomeric equilibria have extremely fast relaxation times. At lower resolutions (between 4.5 and 1.5 Å), the positions of some protons can be deduced from the way hydrogen bonds are organized. Molecular dynamics also provides elements of analysis (Auffinger and Hashem 2007).

5.1.3. *The example of the glmS ribozyme*

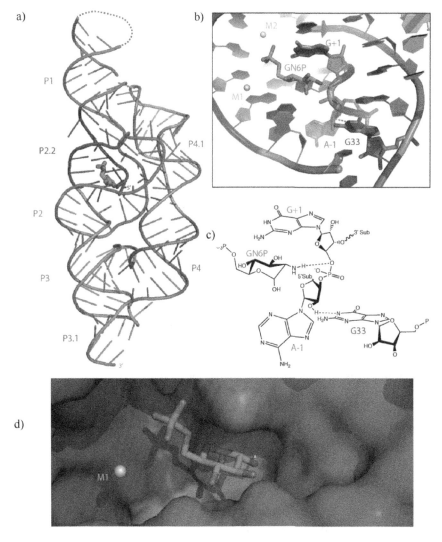

Figure 5.4. *The glmS riboswitch is also a ribozyme and uses glucosamine-6-phosphate (GN6P) as a reaction cofactor*

COMMENTARY ON FIGURE 5.4.– *a) The complexity of the three-dimensional structure (pdb 2nz4 (Cochrane et al. 2007)) results from the tertiary interactions P1/P4.1, on the one hand, and the two pseudoknots P2/P2.2 and P3/P3.1, on the other hand. The two magenta loops connected to P2 are essential for the formation of the catalytic site and thus for the recognition of the GN6P cofactor. b) A detailed view of the catalytic site shows how the global architecture of the ribozyme allows the bases and sugars of the A_{-1} and G_{+1} nucleotides to be moved away from each other to promote the alignment of the O2', P and O5' atoms involved in the S_N2 reaction. This figure shows the hydrogen bonding (yellow hatched line) between G33 (magenta) and the O2' group of A_{-1} whose nucleophilic activated form attacks the phosphate group. The amine of GN6P restores the missing proton of the O5' group of G_{+1} during hydrolysis. The proximity between these groups is materialized by a hydrogen bond (yellow hatched line) set up even before catalysis. c) This scheme is an interpretation of the crystallographic structure and presents the phosphate group interacting with the amine of GN6P while the O2' of A_{-1} interacts with the N1 group of G33. This pattern is very close to the transition state. d) The GN6P ligand binds in a suitable cavity. Two metal ions neutralize the charge of the phosphate; hydrogen bonds are formed between the neighboring bases and the hydroxyl groups of the ligand. G_{+1} interacts by stacking on the ligand, thereby stabilizing the particular conformation favoring S_N2.*

Another ribozyme shows interesting reactivity because it is a metabolic ligand that acts as the stabilizing acid of the leaving group. This is the glmS riboswitch that is located in the messenger regulatory region of glucosamine-6-phosphate synthase (Winkler et al. 2004).

In the presence of low amounts of glucosamine-6-phosphate (GN6P), the ribozyme is inactive and the messenger is translated to produce the enzyme that synthesizes GN6P, which is a precursor for wall synthesis in bacteria and fungi. When the concentration of GN6P increases, it binds to the riboswitch and induces catalysis (Figure 5.4).

Cleavage of the riboswitch generates a 5' hydroxyl end that is a degradation signal for RNase J1 (Collins et al. 2007). The structures of the holo- and aporibozymes are identical, indicating that the regulation aspects are related to catalytic properties.

5.2. An example of catalysis control: lariat-capping ribozyme

We have just illustrated how ribozyme structures organize the establishment of a catalytic core. An essential property of ribozymes is also that they control catalysis by activation mechanisms. In the simplest cases, the control is transcriptional, i.e. as soon as the RNA chain required for folding is transcribed, the ribozyme becomes capable of cutting. In other cases, activation may be dependent on ligand binding. This is the case of the glmS ribozyme that recognizes the glucosamine-6-phosphate or of group I introns whose splicing catalysis relies on the recognition of an exogenous guanosine cofactor. Other ribozymes have undergone conformational adaptations to trigger their catalytic reaction in response to environmental changes. This is the case of the lariat-capping (LC) ribozyme (Johansen and Vogt 1994), which we now discuss.

The LC ribozyme presents an interesting case of functional adaptation. This ribozyme catalyzes the formation of a three-nucleotide RNA lariat following a catalytic mechanism close to that of group II introns (Nielsen et al. 2005).

Like group II introns, it is the O2' group of a nucleotide internal to the ribozyme (U232) that attacks the phosphate group located two nucleotides upstream (Figure 5.5(a)). The original name of this ribozyme GIR1 (Group I-like ribozyme 1) reflects its structural relatedness to group I introns (Johansen and Vogt 1994). The LC ribozyme has so far only been observed in a few unicellular eukaryotes of the genera *Didymium* and *Naegleria* (Johansen et al. 2002) where it is always inserted in the group I intron of the small subunit (SSU) ribosomal RNA coding gene (Figure 5.5(b)).

Catalysis by group I and II introns leads to their own splicing and simultaneous ligation of the flanking exons. LC ribozyme catalyzes a reaction that resembles the first step of the group II intron mechanism that results in the branching of the RNA chain as a three-nucleotide lariat. If the activity of the LC ribozyme were uncoupled from that of the intron into which it is inserted, the ribosomal RNA exons would systematically separate during transcription.

These considerations explain in part why known ribozymes LC copies are inserted into group I introns (Figure 5.5(c)). The LC ribozyme preforms the lariat and thus promotes catalysis, which is triggered only when splicing by

92   Looking at Ribozymes

the host group I intron has occurred (Figure 5.5(d)). The interactions that promote the formation of the lariat are shown on the secondary structure deduced from the crystallographic structure (Figure 5.5(e)).

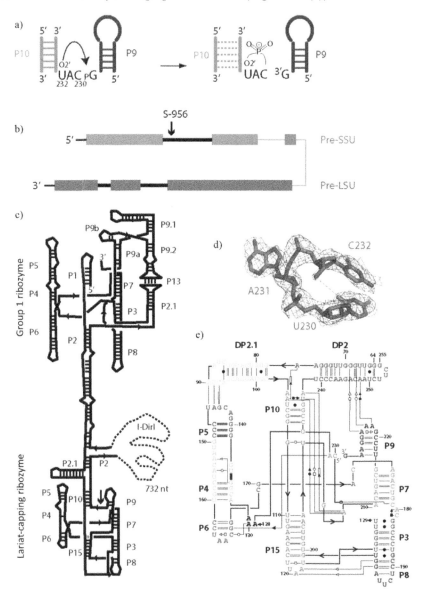

Figure 5.5. *Genetic environment, function and structure of the LC ribozyme*

COMMENTARY ON FIGURE 5.5.– *a)* Didymium iridis *LC ribozyme catalyzes the branching reaction that results in the formation of a three-nucleotide lariat. The O2' group of U232 attacks the phosphate of C230 following an S_N2 mechanism. b) A schematic of the organization of ribosomal DNA is shown. The two introns (black lines) present in the large ribosomal subunit (LSU) gene are classical. In contrast, the intron inserted at position 956 of the small ribosomal subunit (SSU) consists of two ribozymes and a 732-nucleotide (C) open reading frame (ORF) encoding a His-Cys box homing endonuclease, I-DirI (Vader et al. 1999). The green region represents the 5,8S RNA gene (Johansen et al. 1992). c) This diagram shows how the lariat-capping ribozyme is inserted into the P2 domain of the group I ribozyme. d) The lariat structure is pre-organized by the LC ribozyme as shown in the crystallographic structure of the ribozyme (Meyer et al. 2014). The distance between the O2' and O5' groups of U232 and C230, respectively, is less than 5 Å, which leaves room to accommodate the absent phosphate moiety of the RNA whose crystallographic structure has been determined (Meyer and Masquida 2014). The lariat structure model is represented in the experimental electron density map shown as a blue mesh envelope. e) The secondary structure deduced from the crystallographic structure shows the tertiary interactions according to the nomenclature of Leontis and Westhof (2001).*

On the contrary, the LC ribozymes are immediately followed by a sequence encoding an endonuclease that recognizes a unique sequence in the host genomic DNA that is located in the small ribosomal subunit gene in which the intron is inserted (Figure 5.5(c)). The intron of *Didymium iridis* Dir.S956-1 is inserted at position 956 of the small ribosomal subunit RNA (Johansen and Haugen 2001). The endonuclease recognizes a sequence of about 40 nucleotides on either side of locus 956 and cleaves the two strands of DNA, inducing repair by recombination with the uncut DNA containing the intron as template (Figure 5.6). Thus, this intron spreads to genomes that do not contain it, following a process referred to as homing (Lambowitz and Belfort 1993; Goddard and Burt 1999; Qu et al. 2014). This process refers to returning to a place of origin, for example a homing pigeon. It is interesting to note that there is emphasis on the bird's ability to return to its point of departure. The process of homing by the endonuclease must be understood in this sense.

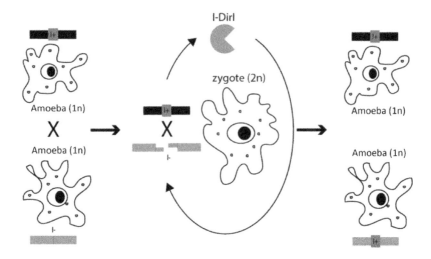

Figure 5.6. *The process of "homing" in* Didymium iridis *occurs during the sexual fusion of two haploid cells (1n-chromosome amoeba), only one of which contains the intron Dir.S956-1 (Johansen and Haugen 2001) to be disseminated (I+ in red in the upper left part of the figure)*

COMMENTARY ON FIGURE 5.6.– *After fusion (center), the endonuclease gene is expressed (blue form). The enzyme I-DirI then cuts the DNA that does not contain the intron at position 956 and the DNA repair by recombination with the DNA that contains the intron that was not cut by I-DirI. The presence of the intron changes the sequence of the target which is therefore protected from the enzyme. The daughter cells (right panel) therefore both contain the I+ intron.*

The LC ribozyme accounts for just 200 nucleotides of the total 1.4 kb of the complete intron consisting of the group I intron into which the LC ribozyme and the coding phase of I-DirI are inserted (Figure 5.5(c)). Its catalytic activity is controlled by the group I intron that harbors it according to the conditions of *D. iridis* life. In the classical pathway, the LC ribozyme is able to "sense" when the intron has spliced. Splicing induces the formation of a structure comprising three contiguous helices involved in the regulatory domain. These three helices form a triple junction DP2/DP2.1/P10 (Figure 5.7). The crystallographic structure of a ribozyme with an activated regulatory domain (Meyer et al. 2014) shows that the two helices (DP2 and DP2.1) upstream of the ribozyme establish tertiary interactions with central regions of the ribozyme that allow communication with the catalytic core

(Figure 5.7). When stabilized by the tertiary interactions, the triple junction forms a specific receptor for an adenine (A209) of a junction important for the initiation of catalysis. This interaction is crucial for the organization of the catalytic core in order to pre-position the residues G208 and A210 that are expected to play the respective roles of base and acid to first activate U232 and then stabilize G229. However, the ribozyme also allows the three residues that form the lariat to be pre-positioned in order to facilitate its closure by transesterification (Figures 5.5(d) and 5.7(e)). Indeed, the O5' and O2' atoms to be linked by the phosphate group are thus within 5 Å of each other. Since this triple junction cannot form until splicing has occurred (Nielsen et al. 2009), the cascade of structural adaptations leading to catalysis is inhibited.

The transesterification reaction produces a three-nucleotide lariat. The reversible ligation reaction occurs in vivo because the phosphate group of C230 – bound to the O2' atom of residue U232 following the formation of the lariat – remains electrophilic and is bound to a stable leaving group. It can therefore be attacked again by the G-terminal residue of ribozyme (ωG). The energy required for the transesterification reaction is retained in the product. This is similar to the case of group I and II introns and endonucleolytic ribozymes. This property of conserving reaction energy explains why ribozymes function without energy input in the form of ATP or GTP. This lariat is found on the 5' of the pre-messenger encoding the I-DirI homing endonuclease and thus forms a special chemical structure that protects the messenger from exonucleases that degrade the RNA from 5' to 3'. This "cap" that extends the chemical integrity of the messenger also stabilizes a hairpin in the 5' end. The latter allows translation of the messenger by the ribosome by a mechanism that is still poorly studied (Krogh et al. 2017).

The crystallographically resolved structure corresponds to the post-catalytic state (Meyer et al. 2014). It provides a good indication of how the regulatory domain participates in the folding of the catalytic core and how it pre-organizes the lariat folding that is formed during the catalytic step. However, the catalytic step itself is not yet understood. This would require solving the initial and intermediate state structures of the LC ribozyme and testing the roles of residues potentially involved in the 208–210 junction.

96 Looking at Ribozymes

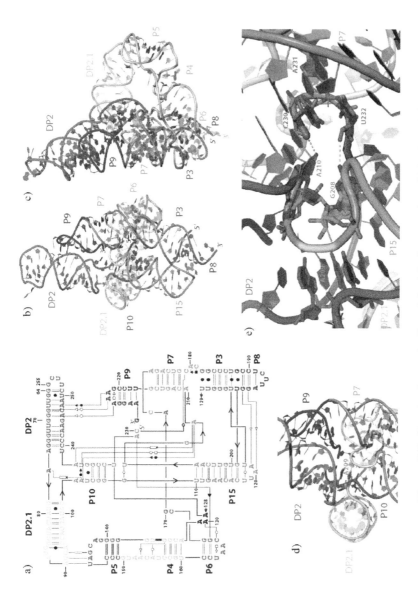

Figure 5.7. *Structural organization of the LC ribozyme from Didymium iridis*

COMMENTARY ON FIGURE 5.7.– *a) The secondary structure is represented taking into account the information deduced from the three-dimensional structure and the color codes are identical. The tertiary interactions interpreted from the 3D structure are indicated according to the Leontis–Westhof nomenclature (Leontis and Westhof 2001). b) and c) The three-dimensional structure (pdb 6gyv) is represented in two different orientations in order to visualize, on the one hand, the tertiary interaction between the loop of P9 and the groove of the DP2 helix (b) and, on the other hand, the pseudoknot formed between the loops of DP2.1 and of P5 (c). The opening of the P8 loop is due to a local disorder in the crystallographic structure. d) The simultaneous formation of the two tertiary interactions mentioned stabilizes the triple junction DP2/DP2.1/P10 that receives A209. e) The Watson–Crick trans base pair between G109 and U207 orients the 208–210 loop favorably for A209 to interact with the triple junction. This gives the loop a structure that allows residues G208 and A210 to point to the three residues forming the lariat and play their respective potential roles as proton acceptors and donors (Meyer et al. 2014).*

The LC ribozymes found in other organisms (*Naegleria* (Wikmark et al. 2006) and *Allovahlkampfia* (Tang et al. 2014)) are inserted into a genetic context similar to *D. iridis*, i.e. within a group I intron and followed by an open reading frame encoding a homing endonuclease. The structure of their catalytic core is conserved, but their regulatory domains formed by P2 and P2.1 are very different from each other and it is not possible to understand the regulatory mechanism by simple transposition of that of the *D. iridis* LC ribozyme. This observation indicates that different adaptive solutions have emerged to address the common problem of coordinating the splicing actions – catalyzed by the group I intron – and lariat formation by the LC ribozyme in these different organisms. These different regulatory domains were probably shaped by different constraints related to the lifestyles of these organisms as well as their species peculiarities.

6

Evolution of the Vision of the Catalytic Mechanisms of Ribozymes, the Hammerhead Ribozyme

Originally, catalytic studies on ribozymes were developed from the knowledge of RNase A at the time. RNase A is able to hydrolyze RNA chains by a mechanism based on the alignment of the nucleophile, the phosphate to be cleaved and the leaving group by distancing the two contiguous nucleotides involved (Raines 1998; Cochrane and Strobel 2008). For this, the nucleophile (O2') must be activated by a base and the enzyme increases the stabilization of the transition state. Finally, the leaving group (O5') is in turn activated by an acid. This chapter traces the evolution of catalytic models of hammerhead ribozymes from its discovery to the present day.

6.1. Chemistry and catalysis: between general acid/base and metal cations

The transesterification reaction catalyzed by hammerhead ribozymes occurs spontaneously under acidic or basic conditions in the absence of an enzyme. It leads to the degradation of RNAs in solution and occurs more so when the RNA is poorly structured as in the case of portions of messenger RNAs (Nielsen 2011). A direct application of this mechanism is in-line probing, which is used to distinguish flexible regions from RNAs of unknown structure. In particular, this method has been used to demonstrate the presence of structured regions in the first identified riboswitches in *Bacillus subtilis* (Regulski and Breaker 2008). The constraints imposed on nucleotides by RNA structuring prevent the appearance of conformations sensitive to spontaneous hydrolysis. A stable structure therefore preserves RNAs from degradation by this mechanism.

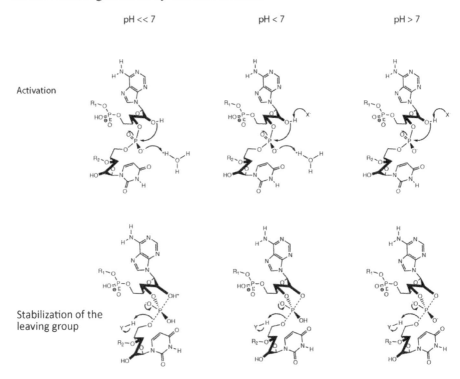

Figure 6.1. *The different types of spontaneous reactions as a function of pH (Emilsson et al. 2003)*

COMMENTARY ON FIGURE 6.1.– *The spontaneous transesterification reaction by the O2' function occurs in the acidic and basic pH range. At very acidic pHs (pH << 7), hydronium ions (H_3O^+) play a dominant role in neutralizing the charges of the phosphate and thus prompting the 2' hydroxyl group to attack it. The situation is totally reversed at pH > 7 where the nucleophile is purely activated by a general base (X). The situation is hybrid at very slightly acidic pH (pH < 7) where the general base and hydronium ions coexist. The leaving group is always stabilized by a proton given by a general acid (Y-H). In the pentavalent oxyphosphorane transition state, the atoms involved in the S_N2 are aligned. The 2' and 5' oxygens are located at the poles of the trigonal bipyramid.*

Under non-enzymatic conditions, the reaction can be catalyzed in acidic or basic media. Compared to the spontaneous reaction, RNase A accelerates the reaction by at least a factor of 10^{11} (Cochrane and Strobel 2008) and only by 10^6–10^9 for hammerhead ribozymes (Canny et al. 2004). The reaction mechanism involves two major chemical steps: (i) nucleophile attack of the activated 2' hydroxyl group of the ribose at position n on the phosphorus of the phosphate group of residue $n+1$ and (ii) breaking of the P–O5' bond releasing the leaving group (Figure 3.7). The two reaction products carry, on the one hand, a cyclic 2'-3'-phosphodiester end (corresponding to the original nucleophile which can then be hydrolyzed to 2'-phosphate or 3'-phosphate), and on the other hand, a 5' hydroxyl end (5'OH) corresponding to the leaving group.

The work of Ronald Breaker provides a better understanding of the situation under non-catalytic conditions where only the pH varies (Li and Breaker 1999; Emilsson et al. 2003). The reaction follows a nucleophilic substitution reaction mechanism of order 2 (S_N2), which is an "in-line" attack where the nucleophile (O2'), phosphorus (P) and leaving group (O5') are aligned. At basic pH, the 2' hydroxyl nucleophile group is activated by deprotonation by a base and becomes an oxyanion (2'-O⁻). The latter attacks the phosphorus of the mono anionic phosphate group (Emilsson et al. 2003; Leclerc and Karplus 2006). In a concerted manner, the formation of the bond between the O2' and the P is accompanied by the breaking of the bond between the P and the O5' of the leaving group. The transition state corresponds to a fleeting intermediate with a characteristic trigonal bipyramid geometry where the phosphorus, the two non-bonding oxygens of the phosphate group and the O3' are in the same plane, while the two bonding oxygens O2' and O5' are located at the two vertices of each pyramid (Scott et al. 2009). The departure of the O5' leaving group is promoted by the transfer of a proton from a water molecule onto the O5' oxyanion (Figure 6.1). At acidic pH, when the phosphate group is neutral or mono-anion, the reaction is activated by protonation of the phosphate group by a hydronium ion (H_3O^+). This activation leads to the attack by the nucleophile in its hydroxyl form. The departure of the leaving group is made possible by the transfer of a proton from a hydronium ion to the oxygen O5' (Taira et al. 1989; Uchimaru et al. 1992, 1993a, 1993b, 1996; Emilsson et al. 2003). At slightly acidic pH, the two mechanisms intermingle with base activation of O2' from the hydroxyl form to the oxyanion, and protonation by a hydronium ion of one of the non-bonding oxygens of the phosphate group.

Another hydronium ion then acts as a proton donor to stabilize the O5' leaving group (Figure 6.1).

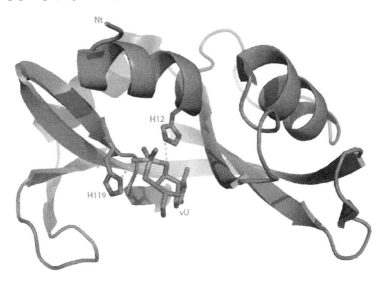

Figure 6.2. *Structure of the catalytic site of bovine ribonuclease A (pdb 6rsa (Borah et al. 1985)). For a color version of this figure, see www.iste.co.uk/masquida/ribozymes.zip*

COMMENTARY ON FIGURE 6.2.– *The two histidines in the catalytic dyad are shown as sticks. Histidine 12 acts as the general base by abstracting the proton from the 2' hydroxyl and histidine 119 stabilizes the leaving group by donating a proton to the 5' oxyanion. The interactions are symbolized by yellow dashed lines. The transition state is represented by a complex of vanadium with a uridine (vU) whose structure was obtained by a combined NMR and neutron diffraction approach.*

In RNase A catalysis, as in other enzymes/ribozymes, several actors are directly involved in the reaction or as cofactors to promote various chemical steps in the reaction. This reaction corresponds to a general acid/base catalysis (Thompson et al. 1995; Schultz et al. 1998). The active site contains two histidine residues. One acts as a general base and activates the nucleophile (H12), the other acts as an acid (H119) and transfers a proton to the 5' oxygen, stimulating the departure of the leaving group (Figure 6.2). Two other residues lysine and phenylalanine (K41 and F120) play a secondary role in catalysis stabilizing the charges of the binding or non-binding oxygens. These residues thus promote the activation of the nucleophile and the departure of the leaving group by acting as a Lewis acid.

Since hammerhead nucleolytic ribozymes generate the same reaction products as RNase A, it was natural to assume that their catalytic mechanism obeyed the same constraints. But the relatively extreme pKa of the chemical functions of nucleotides make them a priori chemically inert at physiological pH (Saenger 1984a). The transposition of the mechanism from RNase A to ribozymes was thus made difficult. The differences between the chemical properties of nucleotides and amino acid side chains led Thomas Steitz (Nobel Prize in Chemistry 2009 for his structure–function studies on the ribosome) to propose a different mechanism based on the action of intermediate chemical species between the biological polymer (RNA or protein) and the substrate. The crystallographic structure of exonuclease-3',5' of *Escherichia coli* polymerase I (Freemont et al. 1988; Beese and Steitz 1991), whose catalytic mechanism is based on the action of two ions Zn^{2+}, Mg^{2+} or Mn^{2+}, illustrates this hypothesis. A catalytic mechanism transposable to the catalytic mechanisms of group I introns and RNase P (Steitz and Steitz 1993) could thus be envisaged.

6.2. Difficulties in interpreting catalysis data

Immediately after their discovery, the hypothesis that ribozymes were metallo-enzymes guided research. Metal cations can act as cofactors for the transfer of protons in general acid/base catalysis and also as Lewis acid/base that capture the free doublets of the oxygen atoms of the RNA groups. Under certain conditions, they promote the appearance of the active form of the nucleophilic 2'OH and the departure of the leaving group by protonation of the O5' by a water molecule belonging to their hydration sphere. Various experimental and theoretical approaches support models of metallo-enzyme catalysis (Steitz and Steitz 1993; Scott et al. 1996; Peracchi et al. 1997; Pontius et al. 1997; Ward and Derose 2012). Experimentally, the studies conducted aim to identify the mode of action of metals and their localization in the ribozyme.

As a Lewis acid, the metal in the first step must facilitate the activation of the nucleophile. Alternatively, the metal can play the role of a general base, a role that is then no longer exclusive since other types of bases can perform this role more efficiently. In this case, the solvated metal behaves like a classical acid/base pair, where the water molecules are the proton donors and acceptors. The pKa of the metal then determines the strength of the base and the reciprocal weakness of the conjugate acid and vice versa. If the metal essentially acts as a Lewis acid, it must be "hard" enough to allow polarization of the O–H bond of the 2' hydroxyl group, without being too "hard" so as not to compromise the nucleophilic character of the 2' oxyanion whose negative charge must attack the phosphate moiety. On the contrary, in the step of breaking the P–O5' bond, the metal has the role of stabilizing the charge of the leaving group on the oxygen O5'. This duality of the role of metals in the two chemical steps of the reaction means that depending on its electronic properties and coordination geometry, a metal will not have an optimal role in the two reaction stages. From the point of view of efficiency as a catalyst, the same analysis applies for general acid/base properties. A strong base and a weak acid conjugate or vice versa cannot a priori be optimal for both chemical steps of the reaction. Depending on the hypothesis to be tested, different metals are studied in a pKa range for general acid/base catalysis or in an electronic charge range for electrostatic catalysis. Moreover, the respective roles of the metals as catalysts are to be put into perspective according to whether the metal is involved in the limiting step of the reaction or not.

In the case of general acid/base catalysis, the hammerhead cleavage reaction is studied in the presence of metals of different pKa (Ca^{2+}, Mn^{2+}, Co^{2+}, Cd^{2+}, Mg^{2+} (Dahm et al. 1993)) to test whether the optimum reaction is dependent on the pKa of the metal tested. Pistol ribozymes were also studied using this method to better understand whether the general acid stabilizing the leaving group was a nucleotide or a metal cation (Wilson et al. 2019). Studies on the hammerhead ribozyme show a correlation between metal pKa and reaction rate as a function of pH, leading the authors to suggest the involvement of a metal-ion hydroxide complex acting as the general base. The interpretation of these data is not straightforward because the pKa indicates the propensity with which a given metal can stabilize a hydroxide ion in coordination, not the availability of this hydroxide ion to act as a base. If the hydroxide ion is strongly bound to the metal, the complex with the metal is a poor base to activate the nucleophile (Lott et al. 1998). The relationship between the pKa of the metal and the pH profile of the reaction may indicate catalysis by a base such as a hydroxide ion specifically activated by the Lewis acid properties of the metal. The first structural data obtained on a crystal-active hammerhead ribozyme (Scott et al. 1996) show several metals, two of which are found relatively close to the cleavage site but in a configuration that is not compatible with a spontaneous reaction. Nevertheless, these mechanisms deduced from the structures of minimal hammerhead ribozymes – too far from the catalytic conformations – have not been confirmed experimentally and have therefore only gone beyond the stage of working hypotheses later invalidated by crystallographic structures integrating the contacts between stems I and II (Figures 5.1 and 5.2) that induce the spacing of the nucleotides between which the cleaved phosphate is located during the reaction (Martick and Scott 2006; Mir and Golden 2016).

The difficulty in interpreting the contributions of cations to catalysis also stems from their very important role in the folding of RNAs and the stabilization of their native structure. In practice, it is therefore often difficult to distinguish "structuring" cations from "catalytic" cations by measures of catalytic activity since a structure trapped in an inactive form cannot catalyze its reaction. Another problem comes from the difficulty in identifying the metals in interaction with RNA in crystallographic structures. Indeed, since X-rays are deflected by the electron cloud of the atoms, small differences in electrons are not easily detected and the nature of the ions cannot necessarily be determined with certainty. For example, the ions Na^+ and Mg^{2+} have the same electronic structure as neon (Ne: $1s^2\ 2s\ 2p^{26}$) and are not differentiable from each other by their diffusion factor. Only their coordination geometry can distinguish them, determined by the number of ligands and their distance from the ion (Auffinger et al. 2011, 2016; Leonarski et al. 2017, 2019). This situation is further complicated by measurement errors in the crystallographic data. RNA crystallographic structures therefore contain many ion assignment errors introducing a strong bias in the interpretation of their catalytic mechanisms (Zheng et al. 2015; Leonarski et al. 2019; Auffinger et al. 2020).

Studies undertaken on the minimal form of hammerhead ribozymes to compare the action of monovalent ions with that of divalent ions suggest a non-specific role for the latter. At high concentrations, monovalent cations play a role identical to that of divalent cations in several nucleolytic ribozymes including hammerhead ribozymes (Murray et al. 1998a, 1998b; Curtis and Bartel 2001; O'Rear et al. 2001). Although the catalytic activity is reduced by more than an order of magnitude, the use of particularly high, non-physiological concentrations of these ions indicates that the reaction path followed under these conditions is the same as in the presence of divalent cations. This assumption suggests that the divalent metals do not intervene as a base in the reaction mechanism.

Historically, the general acid–base catalysis model involves one metal cation, while the electrostatic catalysis model involves two metal cations (Dahm et al. 1993; Pontius et al. 1997; Lott et al. 1998). In the one-metal cation mechanism, the metal activates the nucleophile but can potentially also facilitate the departure of the leaving group by acting as a general acid through proton transfer to the O5' oxygen. The ambivalent role of a single metal cation is only appropriate in an asynchronous two-step mechanism. In the two metal cation mechanism, the presence of two metals leads to a priori a concerted reaction between the two steps of the reaction. In either type of catalysis (general acid–base or electrostatic Lewis acid), there is no data with a direct measurement of the positions of the metals or their exact role in the catalysis. The questions that arise are therefore whether the metals are involved: (1) as catalysts with an involvement in proton transfers or (2) as cofactors to stabilize intermediates or transition states without involvement in the chemical steps of the reaction. To elucidate the role of metal cations and the type of associated catalysis, several experimental approaches are used. For this purpose, the different affinity of different metal cations for phosphorothioate-modified nucleotides is exploited. Deuterated water (D_2O) is used as a solvent in order to identify the limiting step of proton transfer by measuring the kinetic isotopic effect, which consists of measuring the variation of the reaction rate when one of the atoms involved is replaced by one of its isotopes.

The principle of using phosphorothioates is based on the differential affinity of "soft" metals such as Mn^{2+} for the oxygens of the phosphate group or their thio analogues where the oxygen is replaced by a sulfur atom. Mn^{2+} binds more favorably to the thio analogues, while a "hard" metal such as Mg^{2+} has a similar affinity for both. However, due to a change in the reactivity of these ribozyme analogues, the expected activity is lower, the so-called "thio effect". If the activity is higher in the presence of Mn^{2+} than in the presence of Mg^{2+}, then it is likely that a metal is bound to the sulfur-substituted oxygen (O2', O3', O5', O1P or O2P). If no significant difference in activity is observed, then the opposite is assumed to be true, that no metal is bound. Since proton transfers are slower in D_2O, an isotopic effect is expected when they occur during the limiting step. The consequences of the thio and isotopic effects on catalysis have been interpreted in favor of either a two-metal cation mechanism or a one-metal cation mechanism.

A thio effect is observed for one of the non-bridging oxygen atoms (O2P in this case) (Figure 1.2a) substituted by a sulfur atom, suggesting that a metal in coordination with this oxygen could intervene in the stabilization of the oxyphosphorane transition state. The lack of a thio effect for the substitution of O5' to sulfur is first interpreted in favor of a one metal cation model rather than a two metal cation model (Kuimelis and McLaughlin 1996). Substitution of O5' with a sulfur leads to a change in reactivity, where departure of the leaving group is spontaneously facilitated in the absence of metal cations. This greater reactivity can be transposed to the ribozyme context (Pontius et al. 1997), especially since it is assumed that the limiting step of the reaction corresponds to this chemical step. In the case of a 5'-sulfur, the limiting step would no longer be the departure of the leaving group (whose reactivity is accelerated by a factor of 10^6 under non-enzymatic conditions) but the nucleophilic attack. Under these conditions, the presence of a metal at 5' becomes superfluous and has no impact on the global reactivity.

Studies on the catalytic mechanism of the minimal form of the hammerhead ribozyme have allowed for original methodological developments diversifying the toolbox of RNA biochemists. However, it should be noted that despite these methods, it was the identification of the complete form of the hammerhead ribozyme (Canny et al. 2004) in which the interaction between stems I and II induces the formation of a pro-active catalytic site that finally elucidated the catalytic mechanism (Wilson et al. 2019). Published results and studies on other nucleolytic ribozymes draw molecules with very different structures but all constrain a similar conformation of their active site. In particular, the two nucleotides surrounding the cleavable phosphate splay apart to the extreme so that the three atoms involved in the S_N2 (O2', P, O5') are best aligned. From one ribozyme to another, however, the strategies for activating the nucleophile (O2') and neutralizing the oxyanion (O5') of the leaving group can vary.

The specific acid/base strategies for each ribozyme are listed in Table 6.1. The general base that activates the 2'-hydroxyl is a guanine, of which imino N1 is mainly used, except in the cases of the hepatitis delta virus (HDV) or twister sister (TS) ribozyme where the acid/base properties of an ion are exploited. The solutions for neutralization of the 5' oxyanion of the leaving group are more diversified. The most classical is represented by the N1 group of adenines or the N3 of cytosines. Studies on the ribozyme HDV have shown that the structural environment of the ribozyme induces the variation of the pKa of the N3 function of the C75 nucleotide (Luptak et al. 2001), stimulating its function as a general acid neutralizing the 5' oxyanion at physiological pH, and thus the rate of the reaction (Nakano et al. 2000; Das and Piccirilli 2005). The pKa change to make the proton of a 2'-hydroxyl function acidic at physiological pH – whose pKa is normally around 13 (Li and Breaker 1999; Emilsson et al. 2003) – can also be caused by interaction with a divalent cation as in hammerhead ribozymes (Wilson et al. 2019). The metal then acts as a Lewis acid by attracting the 2' oxygen doublets to itself in a way that makes its proton more labile. In contrast, a divalent hydrated cation ($Mg^{2+}(H_2O)_6$) can directly neutralize the 5' oxyanion by donating a proton from one of its ligand water molecules. In the case of the pistol ribozyme, the active site ideally positions the cation relative to the phosphate group to be cleaved, leading to protonation which neutralizes the 5' oxyanion of the leaving group.

Ribozyme	General base	General acid
Hammerhead (Wilson et al. 2019)	G	G O2' – Me or Me
Hairpin (Rupert et al. 2002)	G	A
VS (Wilson et al. 2007; Suslov et al. 2015)	G	A
Twister (Wilson et al. 2016)	G	A
HDV (Das and Piccirilli 2005; Chen et al. 2010a)	$Me^{2+}OH_2$	C
Twister sister (Liu et al. 2017)	$Me^{2+}OH_2$	C
Pistol (Wilson et al. 2019)	G	Me^{2+}
GlmS (Cochrane et al. 2007)	G	GN6P
Hatchet (Zheng et al. 2019)	G	G O2

Table 6.1. *Strategies for activation of the O2' nucleophile (general base) and stabilization of the 5' oxyanion of the leaving group (general acid) (from Lilley (2017))*

COMMENTARY ON TABLE 6.1.– *The N1 of guanines is used except for the hatchet ribozyme where the N7 is optimally positioned. In the case of acid, the N1 of adenines or the N3 of cytosines is involved. Ions such as Mg^{2+} can also be used instead of nucleotides.*

An atypical strategy for protonation of the 5' oxyanion is observed in the case of the GlmS ribozyme where the ligand of this ribozyme – which is also a riboswitch – glucosamine-6-phosphate (GN6P) is the proton donor. Thus, the ribozyme is active only when the ligand reaches the concentration that indicates the need for translation arrest of the enzyme glucosamine-6-phosphate synthase (Winkler et al. 2004), which is the enzyme producing GN6P. Thus, the product of the enzyme directly regulates its own production (Figure 5.4). Nevertheless, the amounts of GN6P accumulated in the cell can be significant since in *Bacillus anthracis*, GlmS has a low affinity for GN6P with a dissociation constant Kd = 1 mM. Since ribozyme is fastest at pH 7.5, the amine moiety of the ligand, with its pKa around 8, is the most likely proton donor (Cochrane et al. 2009).

To date, GlmS and group I introns are the only ribozymes whose catalytic strategies rely on the use of an exogenous organic ligand. In each case, the ligands are derived from sugars (αGTP contains ribose and GN6P glucose) and are used in different steps of the catalytic mechanism. In group I introns, αGTP performs the nucleophilic attack following its activation by a Mg^{2+} ion (Figure 2.4). In the case of GlmS, it is the last chemical step that is targeted, the protonation of the leaving group by a weak acid, in this case the protonated amine of GN6P. The rarity of the participation of exogenous ligands in ribozyme reactions is remarkable. This mode of positive or negative feedback control of activity is frequently encountered in proteins and is extremely simple and sophisticated. It is therefore reasonable to hypothesize that as yet unknown biological pathways relying on ribozymes depend on this type of control.

7

The Distribution of Ribozymes in Living Organisms and Molecular Adaptations during Evolution

7.1. Ubiquitous ribozymes

The ribozymes corresponding to group I/II introns and RNase P are distributed throughout the living kingdom (Figure 7.1). Nevertheless, the selection pressures that lead to their dissemination are of a different nature. Self-splicing introns behave as mobile elements capable of horizontal transfer. Their appearance at various locations in the genome is therefore neutral a priori. However, the integration of these mobile elements gives the host a selective advantage by conferring a specific sequence to the affected genes that prevents the occurrence of recombination accidents when cells are exposed to exogenous homologous DNAs (Lambowitz and Zimmerly 2004; Lambowitz and Belfort 2015).

Group I introns are also found in bacteria, but their frequency is much higher in plants and unicellular eukaryotes such as amoeba, fungi and yeast (Haugen et al. 2005a).

Group II introns, meanwhile, are found in the organelle genomes of these organisms and also in bacteria (Bonen and Vogel 2001; Bonen 2012).

For a color version of all the figures in this chapter, see www.iste.co.uk/masquida/ribozymes.zip.

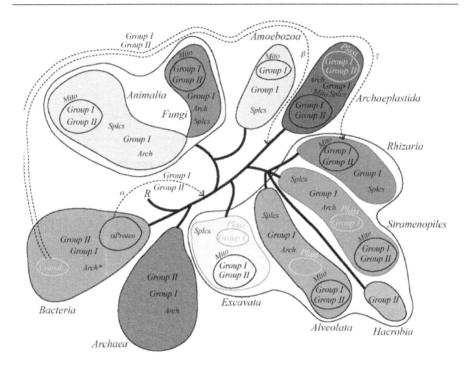

Figure 7.1. *Phylogenetic tree showing the distribution by phylum and organelle of group I and II introns, spliceosomal (Splcs) and archaeal (Arch) introns (Rogers 2019)*

COMMENTARY ON FIGURE 7.1 – *The tree aggregates information from several sources (Baldauf 2003; Haugen et al. 2005a; Roger and Simpson 2009; Hug et al. 2016). The tree (with R as the root) in solid lines indicates vertical transmission of group I and II ribozymes. The hatched lines indicate horizontal transmission events of ribozymes. The presence of ribozymes in mitochondria (Mito) and plastids (Plaste) is also indicated. This results from the formation of mitochondria from α-proteobacteria, an event that occurred after the division of archaea and eukaryotes (α). Two other sources of ribozymes originate from a cyanobacterium that formed chloroplasts in archaea (β) or cyanelles in rhizoids (γ). The regrouping of the phyla of Excavata, Alveolata, Hacrobia, Stramenopiles and Rhizaria indicates that these organisms are single-celled eukaryotes. Archaeal introns often* *localized in pre-tRNA or pre-rRNA are spliced by tRNA-specific endonucleases (Tocchini-Valentini et al. 2011). Archaea-like introns are also found in bacteria (Arch*).*

The ubiquitous tRNA-maturation function of RNase P directs selection pressure to maintain conserved forms of ribozymes from very few ancestral organisms. The selection pressure therefore directs its maintenance and fidelity of activity and low divergence from its homologues. The recent discovery of a unique RNase P consisting of protein subunits in human and plant mitochondria (Holzmann et al. 2008) also indicates that the boundary between an RNA world and a protein world is tenuous since these two distinct polymers are both capable of performing the same reactions on tRNA substrates that they specifically recognize. Nucleolytic ribozymes, on the other hand, have a well-defined functional role in viroids, virusoids and other viruses for the replication of their genome. But as we mentioned at the end of the previous chapter, we will see later that their presence is not limited to these specific niches of the living world.

7.2. Selection pressures at work in ribozyme shaping

As with proteins, the sequences of RNA molecules adapt to environmental, physiological and metabolic selection pressures. These adaptations result from the selection of sequences undergoing natural mutations (sequence drift) (Ohno 1970; Kimura and Ohta 1974; Bershtein Tawfik 2008) and/or from deletion or insertion events following the duplication of certain genes in a loss/gain of function cycle. In the simplest case, gene duplication partially frees the duplicated gene from the selection pressure applying to the original gene product, which continues to exercise its function.

Two main types of selection apply to duplicated genes. Either the function is retained in an altered biological context (adaptive selection) or a new function emerges (negative selection) (Smith et al. 2013). It is in the untranslated regions of genomes that negative selection primarily operates. RNAs generated from these transcriptional islets thus constitute a pool of diversity. They may have a function (lncRNA, sncRNA, siRNA, ribozymes) or not, or they may be in the process of acquiring a function following mutation events related to the maintenance of the genome, its replication and transcription.

In this section, we describe examples of these molecular adaptations at different scales of detail. First, adaptive selection will be discussed in cases of equivalence between structural motifs and then we will show how the same function can be maintained in a different structural environment. Finally, we will present a case of negative selection where the abrupt change in the nature of the selection pressures allows the release of sequence drift in a way that favors the emergence of a new function.

An example of adaptive evolution involving a well-studied RNA is the hammerhead ribozyme. In this case, the maintenance of tertiary interactions between helix I and II loops is critical for high activity under physiological conditions. Contacts between helices are held by structural deformations induced by internal loops that do not form Watson–Crick *cis*-pairings. This impossibility orients the nucleobases outward from the loops, which thus establish contacts between stems I and II (Figures 5.1 and 5.2). The position of the strain relative to the first base pair of each helix is therefore important in allowing these nucleotides to interact. The crystallographic structure of the *Schistosoma mansoni* hammerhead ribozyme illustrates how such a set of interactions arises (Martick and Scott 2006). Although exemplary structures of this ribozyme belonging to other species have not been resolved, comparative sequence analysis and catalytic assays on ribozymes in which these contacts have been altered by mutagenesis show their maintenance (Khvorova et al. 2003). These observations illustrate that sequence drift samples the space of pairing possibilities in such a way as to present multiple structural solutions responding to the same selection pressure. The numerous genomic copies of hammerhead ribozymes illustrate this concept (Hammann et al. 2012). Regardless of which stem the hammerhead ribozyme is connected to in the genome, RNA always produces an interaction between stems I and II.

A classic example of adaptive selection for large (i.e. longer sequence) RNAs is represented by ribosomal RNAs whose many genomic copies generally evolve in a concerted fashion. However, many exceptions to this rule have been described. Some ribosomal RNAs may have lost/gained some functionality as a result of mutations, insertions or deletions. Eukaryotic ribosomes have additional RNA segments compared to prokaryotic counterparts. These expansion segments confer specific functions (Fujii et al. 2018). Although the majority of these functions are unknown, the 159-nucleotide ES27L expansion inserted into domain IV (helix 62) of yeast 25S RNA appears to be involved in increasing translation fidelity (Fujii et al.

2018). Furthermore, the coexistence of ribosomal RNA copies with slightly different sequences allows these pseudogenes to respond to existing environmental selection pressures or their modifications, thus conferring adaptability to the organism. A striking example is the mitochondrial ribosome, which has lost large regions of RNA that have been replaced by proteins that functionally complement these lost domains. The 5S RNA, which participates in the constitution of the large subunit, is itself replaced by a mitochondrial tRNA. This replacement is specific, a phenylalanine tRNA in *Sus scrofa* (Greber et al. 2014, 2015) and a valine tRNA in humans (Amunts et al. 2014).

This example is also found in group I and II introns and RNase P and generally in ribozymes. Bacterial RNase P is represented by two subtypes, A and B, in which distinctive elements of the secondary structure have equivalent roles in the three-dimensional structure (Haas et al. 1996). Each ribozyme family exhibits heterogeneities in peripheral domains across organisms. Nevertheless, the structure of the catalytic core is maintained and thus retains the function, two-step splicing (Lehnert et al. 1996) or cleavage of the 5'-leader end of tRNAs (Massire et al. 1998). This is adaptive evolution. Molecular adaptability can go even further, however, since RNase P and autocatalytic introns are found alongside molecules of distinct nature and structure in the cell that are capable of performing the same catalytic actions. Notably, plant mitochondria contain a protein RNase P (Holzmann et al. 2008) and yeast have a spliceosome (Wilkinson et al. 2020), while mitochondrial pre-messenger splicing is performed by 11 group I and II introns (Bonen and Vogel 2001; Turk et al. 2013). Although coexisting in yeast *Saccharomyces cerevisiae* mitochondria, these group I and II introns all have different secondary structures and regulatory mechanisms. These differences allow yeast to regulate pre-messenger splicing in response to environmental variations.

Known examples of negative (or purifying) RNA evolution are much rarer. Among natural ribozymes, the archetype is the LC ribozyme (Nielsen et al. 2005). Its structure and genomic location suggest that its ancestor was likely a group I intron. The LC ribozyme is systematically inserted in a group I intron and therefore functions synergistically with it (Vader et al. 2002). The structural relationship with a group I intron and the position of the LC ribozyme in the intron create a mechanism for its appearance (Figure 7.2). Initially, a classical group I intron would have inserted by reverse splicing into a group I intron present in the small ribosomal subunit gene, into which a sequence encoding a specific endonuclease was inserted (Birgisdottir et al. 2011). This localization would have allowed the newly inserted intron to escape the selection pressures associated with splicing, facilitating the drift of its sequence. This drift would have led gradually (we are talking in millions of years) to the formation of the P7/P15 pseudoknot in addition to the one formed between P3 and P7, to the shortening of the peripheral structural elements and to the selection of the reaction leading to the formation of the three nucleotide lariat (Beckert et al. 2008). The formation of the lariat conferring the endonuclease-encoding messenger carried by the ancestral intron a substantially longer lifespan would thus have been maintained during the evolution of this nested ribozyme system.

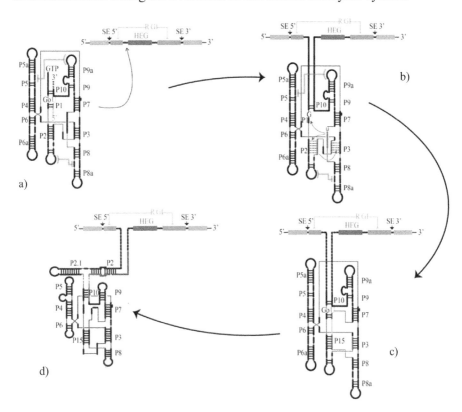

Figure 7.2. *Model of the events that may have led to the appearance of the LC ribozyme*

COMMENTARY ON FIGURE 7.2.– *The first event (a) certainly involved the insertion of a group I intron (schematized by its secondary structure P1–P10) into another group I intron (green boxes, Group I Ribozyme; R GI) established in the small ribosomal subunit RNA gene (orange boxes) by reverse splicing (Birgisdottir and Johansen 2005). In this very particular context, the inserted intron was removed from the selection pressures of splicing since R-GI retained its function. R-GI contains an open reading frame encoding an endonuclease (HEG: Homing Endonuclease Gene, red box) inserted by homing (see Figure 5.6) at position 956 of the gene (Haugen et al. 2005b). Once inserted in a splice-neutral region of R-GI but upstream of the 5' end of the endonuclease, the newly inserted group I intron is subject to different selection pressures (b). The latter can therefore be mutated without affecting ribosomal RNA maturation. Mutations have thus appeared gradually leading to the formation of the P3/P7/P15 double pseudoknot (c). Other mutations have led to the appearance of the branching reaction and its regulation in a concerted way with splicing by R-GI (d). The gain in stability brought to the messenger of HEG by the three-nucleotide lariat formed in the 5' and acting as an alternative cap promoted the maintenance of the ribozyme within the intron. The rarity of examples of this type of intron indicates that ribozyme nesting results from low-probability events and that the mutations necessary for their function to emerge occur over a time scale of millions of years (Goddard and Burt 1999).*

7.3. Ribozymes in cellular processes: from viroids to eukaryotes

Within the family of autonucleolytic ribozymes, hammerhead ribozymes were the first to be identified and the most frequently encountered to date (Hammann et al. 2012; Maurel et al. 2019). Their initial discovery took place in viroids (see Chapter 3), the smallest known pathogens, infecting plants and also mammals. The viroid genome is composed of a circular single-stranded RNA molecule of up to 400 nucleotides that does not encode any proteins. The viroid genome is therefore a circular, non-enveloped, non-coding RNA with no enzymatic activity of its own, such as that usually found in viruses (polymerase, integrase, protease, etc.). These pathogens are considered as "living" fossils that replicate within their host thanks to three enzymatic activities of polymerase, ligase and nuclease. From a phylogenetic point of view, viroids are grouped into two families, the *Avsunviroidae* (Avocado Sunblotch Viroid) and the *Pospiviroidae* (Potato Spindle Tuber Viroid), whose genome replicates, respectively, in the chloroplast or in the cell nucleus of the host plant according to the mechanism known as the rolling circle mechanism (Flores et al. 2009). In addition to this difference in intracellular localization, the mode of replication is said to be symmetric in the case of the Avsunviroidae because the [+] and [-] strands are replicated in an equivalent manner (Figure 3.8). In *Pospiviroidae*, the mechanism is said to be asymmetric. Transcription of the circular [+] strand produces a linear [-] strand composed of several copies (concatamer). This concatameric [-] strand in turn serves as a transcriptional template to generate new copies of the [+] strand before the circular genomes are formed. In *Pospiviroidae*, all three enzymatic activities required for viroid replication are host-derived, for example RNase III for the nuclease activity in the plant. In *Avsunviroidae*, the polymerase and ligase activities also come from the host, but the nuclease activity is carried by the viroid in the form of a hammerhead RNA motif that allows self-cleavage of the replicated genome (Flores et al. 2009; Cervera et al. 2016). Another significant difference between the two viroid families concerns the chemical nature of the cleaved strand ends. In *Pospiviroidae*, RNase III produces a 5' phosphate end and a 3' hydroxyl end. A ligase is therefore necessarily involved in circularizing individual genomes. In contrast, in *Avsunviroidae*, the ribozyme generates the 5' hydroxyl and 2',3' cyclic phosphodiester ends. The action of a ligase is possible but not required, since the binding chemical energy stored in the phosphodiester form is usable by the ribozyme.

Comparative analysis of hammerhead ribozyme sequences identified in the genomes of different viroids of the family *Avsunviroidae* (Hutchins et al. 1986; Forster et al. 1987) identified conserved residues essential for catalytic activity. This model is proven and refined by experiments to determine the minimum RNA configuration that retains catalytic properties and results in the definition of a minimum motif size of less than 50 nucleotides (Uhlenbeck 1987).

Subsequently, many hammerhead motifs have been discovered in all of the living kingdom by in silico approaches. It would be too long to describe here the thousands of examples found, but it is possible to imagine mechanisms by looking at the genetic contexts in which these ribozymes are inserted (Figure 7.3).

a) tRNA maturation

b) Production of regulatory RNAs

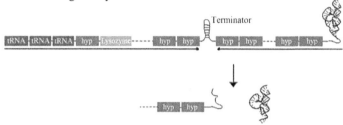

c) mRNA maturation

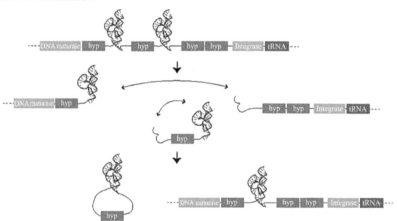

d) Inhibition of RNA interference

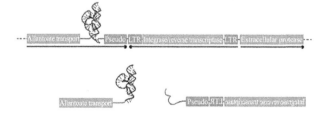

Figure 7.3. *Potential mechanisms of hammerhead ribozymes based on their genetic context (Hammann et al. 2012)*

COMMENTARY ON FIGURE 7.3.– *(a) Ribozymes can participate in the maturation of polycistronic pre-tRNAs as in this example from a marine metagenome. (b) They can also serve to produce regulatory non-coding RNAs. A link is evoked with the inhibition of expression from a* Pseudomonas aeruginosa *promoter. This mechanism seems to be involved in the integration of the PaP3 phage DNA in this bacterium. (c) In* Azorhizobium caulinodans*, ribozyme activity could lead to the rearrangement of exons and thus generate new forms of messengers, including circular forms. This activity is similar to the alternative slicing orchestrated by introns in eukaryotes. (d) In the fungus* Yarrowia lipolytica*, the ribozyme action could prevent the synthesis of two complementary strands that could result from the initiation of transcription from two promoters of opposite directions (transposon LTR (Long Terminal Repeat) promoter, mRNA encoding the allantoate transporter). This mechanism avoids messenger degradation by the RNA interference mechanism.*

Hammerhead ribozymes are notably present in the human, broader mammalian, reptile and bird genomes (Hammann et al. 2012). According to Hammann et al. (2012), hammerhead ribozymes that punctuate polycistronic tRNA pretranscripts could initiate their maturation. In bacteria such as *Pseudomonas*, these ribozymes can serve to generate small non-coding RNAs. Tandems of hammerhead ribozymes in pre-messengers can also lead to splicing of a coding part and thus mimic the alternative splicing phenomena that result in the synthesis of proteins with truncated domains. The religation of ribozymes can even generate circular RNAs with exons. Ribozyme cleavage can also prevent the generation of double-stranded RNAs when promoters are located on opposite sides of the same DNA locus; preventing the synthesis of the complementary strand results in the inhibition of RNA interference mechanisms (Fire 2007) so as not to repress the translation of a given protein.

In addition to these scenarios, there are better characterized situations such as transposons, genetic elements capable of transposing from one part to the other of the genome according to well established mechanisms. The transposons are classified into different families according to their mechanism of transposition. When an RNA intermediate is used in the transposition mechanism, it is called a retrotransposon (Aizawa et al. 2003). These elements constitute about 40% of the human genome (Li et al. 2001).

In one retrotransposon family, hammerhead ribozymes are actively involved in the mechanism (Lunse et al. 2017). In other retrotransposon families, it is the HDV ribozymes that are more represented (Weinberg 2021). The hammerhead ribozyme retrotransposon family constitutes the Penelope elements (Evgen'ev and Arkhipova 2005) – the names of several transposable elements come from Greek mythology and refer to the life of Ulysses in Homer's *Iliad* and *Odyssey* (Homer 1992, 2003), Penelope, Telemachus, Paris and Helen (Evgen'ev et al. 1997; Evgen'ev 2013). These elements encode an enzyme consisting of a reverse transcriptase domain followed by an endonuclease domain (Evgen'ev 2013). Reverse transcription synthesizes complementary DNA integrated into the genome from the RNA naturally transcribed by the cellular polymerase. The endonuclease cuts the DNA at loci rich in AT residues and allows repair of the cut from the complementary DNA of the neosynthesized RNA (Gladyshev and Arkhipova 2007). This is not unlike the homing process described in Figure 5.6. Penelope elements are flanked by two hammerhead ribozymes attached to the genome via their stem I. The III stem of these ribozymes is usually short or even non-existent, which strongly compromises their phosphodiesterase activity (Cervera and De la Pena 2014).

In order to cut efficiently, a dimer of ribozymes is formed, which allows for the generation of a true stem III (Figure 7.4) that enables the folding and activity of the catalytic core (Leclerc et al. 2016; Lunse et al. 2017; Maurel et al. 2019). However, dimerization of these ribozymes is only observed in in vitro catalytic assays.

Nevertheless, two examples of ribozymes active in dimeric form have already been observed by structural approaches. These are the VS ribozyme (Saville and Collins 1990), already discussed in the history of catalytic RNA discoveries section of this book (see Figure 3.9(d)) and the hatchet ribozyme (Weinberg et al. 2015; Zheng et al. 2019) (see Figures 3.10(e) and 7.4(b)).

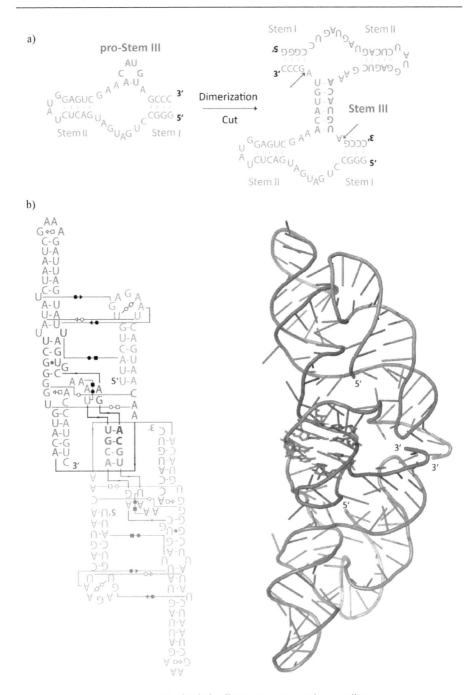

Figure 7.4. *Nucleolytic ribozymes are active as dimers*

COMMENTARY ON FIGURE 7.4.– *(a) Dimerization of hammerhead ribozymes gives rise to a stem III that enables catalytically active ribozymes to be formed (from Lunse et al. (2017)). The locations of the cuts are marked with black arrows. The colors of the ribozymes help distinguish each monomer. Stems I are formed by hybridization of the complementary regions between the two ribozymes. (b) Representation of the dimerization observed in hatchet ribozyme crystals (pdb: 6jq5). The second copy of the ribozyme is colored uniformly in gray for simplification. The way the 3' ends of each ribozyme (in green and gray) twist around each other is clearly visible. This structure suggests that ribozyme activity may be controlled by dimerization as is the case for ribozyme VS in which the active site is formed in a composite fashion by the addition of nucleotides essential for catalysis from each of the monomers (Suslov et al. 2015).*

7.4. Very human ribozymes

Hammerhead ribozymes are also found in humans and mammals. Marcos de la Pena's team identified some of them in introns (de la Pena and Garcia-Robles 2010). In the case of the RECK protein, a tumor suppressor (Takahashi et al. 1998), a ribozyme is found in the middle of the sixth intron. In another example, the ribozyme is found in the first intron of the CTCL (cutaneous T-cell lymphoma) tumor antigen gene (Hartmann et al. 2004). Another hammerhead ribozyme is also found in the first intron of the pre-messenger of another protein, DTNB (Dystrobrevin beta protein) in sauropsids such as lizards or birds. These locations probably indicate a possible role in splicing. But other examples report the presence of ribozymes in the 3' untranslated ends (3'-UTR) of lectins from mammals (Martick et al. 2008a; Scott et al. 2009). Their presence in transposable elements indicates that they disseminate vertically (from one cell to its progeny) or horizontally (from one organism to another via inter environmental actions) and reinforce the link between the RNA world and the origins of life (Hammann et al. 2012; Lunse et al. 2017).

As already mentioned, ribozymes are found throughout the living kingdom. Many transposable elements also contain group I or II introns that also code for proteins that enable reverse transcription or endonucleolytic cleavage at specific sites (Aizawa et al. 2003). On the contrary, 90% of group I introns are mainly present in land plants, red (rodophytes) or green (chlorophyta) algae, fungi and yeasts. They are also present in bacterial genomes to a lesser extent (see Figure 7.1). They can be located in nuclear DNA, in the genomes of organelles such as plastids or mitochondria (Haugen et al. 2005a). Group II introns are present in bacteria (25% of sequenced bacterial genomes contain at least one group II intron), and in the organelle genomes of yeast, fungi and plants. No group II introns are present in eukaryotic nuclear genomes, and few have been identified in archaea (Lambowitz and Belfort 2015).

A new family of ribozymes has recently been identified in long known retrotransposons, the SINEs (Short Interspersed Nuclear Elements) (Hernandez et al. 2020). The authors of this work indicate the important characteristics of these interspaced short repeated nuclear elements. They constitute a class of retrotransposons like LINEs (Long Interspersed Nuclear Elements) or LTRs (Long Terminal Repeats). SINEs correspond to transpositions of polymerase III transcripts, thus of non-coding RNA messengers such as tRNAs, 5S ribosomal RNA and spliceosomal RNAs. These are the Alu sequences in humans whose name derives from the cleavage sites by the Alu-I endonuclease (Baillie et al. 2011). These transpositions rely on the reverse transcriptase and endonuclease activities of proteins encoded by LINEs (Deininger 2011). SINEs account for approximately 11% of the human and murine genomes. In mice, phylogeny subdivides their homologues into four distinct families, B1, ID, B4 and B2, which alone accounts for about 4% of the genome (Chinwalla et al. 2002). In humans, the Alu elements are divided into three subfamilies, J, S and Y (Deininger 2011). These RNAs (B2 and Alu) slow down polymerase II (which transcribes genes yielding protein messenger RNAs in particular) by hybridizing to the chromatin of stress response genes (heat shock), thus inducing pauses during transcription. If a stress occurs, the Polycomb protein (EZH2) is recruited to B2 or Alu RNAs to induce ribozyme cleavage. Once the RNA is cut, the released genes are quickly transcribed triggering an instant stress response. It was after unraveling this complex mechanism that Jeannie Lee's lab at Harvard realized that the interaction with Polycomb allowed B2 and Alu RNAs to fold into a structure capable of self-cleaving and then self-degrading leaving polymerase II free to transcribe stress genes without constraint (Hernandez et al. 2020). The B2 and Alu ribozymes can therefore be considered as epigenetic ribozymes since they are able to respond to the action of Polycomb, which is known to have histone H3 methylation activity (Margueron and Reinberg 2011). For reasons of comparison, the ribozymes contributed just by SINEs account for 4% of the genome that must be set against the 2% of sequences encoding proteins. This makes it possible to realize how undervalued the role of catalytic RNAs is in current research.

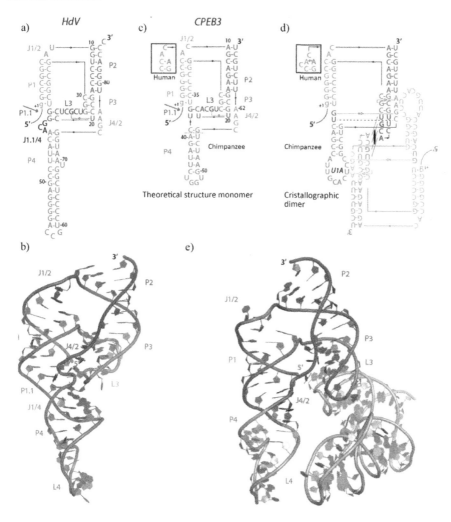

Figure 7.5. *Hepatitis delta virus (HDV) ribozyme and a related human ribozyme identified in the CPEB3 protein gene*

COMMENTARY ON FIGURE 7.5.– *(a) Secondary structure of the HDV ribozyme deduced from the crystallographic structure. (b) ((Ferré-D'Amaré et al. 1998), pdb: 1drz) showing the two pseudoknots required for catalysis (P1/P2 and P3/P1.1). (c) The related human ribozyme is the one found in the second intron of the CPEB3 protein gene whose secondary structure extrapolated from that of the HDV ribozyme is shown. (d) The crystallographic structure indicates, however, that the P3/P1.1 pseudoknot is not formed, giving a different secondary structure in which the P3 loop is released. Four nucleotides of this palindromic 5'-ACGU-3' sequence loop pair with corresponding nucleotides of a second ribozyme linked by a second-order axis of symmetry (black elliptical symbol at the 4 bp forming the dimer) in the crystal. (e) As can be seen in the crystallographic structure (pdb: 7qr4), this 4 bp helix sequesters the nucleotides that should form P1.1, which confers to the ribozyme of CPEB3 a reduced catalytic performance compared to the ribozyme of HDV. The identical color coding of the structural elements allows visualization of the differences between the three secondary structure diagrams.*

Hundreds of hairpin ribozymes have been found in sequences of environmental metatranscriptomes (Weinberg et al. 2021). However, these ribozymes leave no trace in genomes. They are therefore ribozymes belonging to biological forms that have an RNA-form genome. These ribozymes probably belong to new families of viruses, viroids and virusoids whose diversity could not be correctly estimated until now. Circular RNAs related to HDV but nevertheless very divergent have been identified in many branches of life. Some of them have a hammerhead ribozyme in place of an HDV ribozyme, notably in metazoans (de la Pena et al. 2021) but retain a coding phase related to the HDV delta antigen.

HDV is the only known human pathogen to contain a ribozyme (Sripathi et al. 2015). In silico research has identified HDV-like ribozyme candidates in many genomes. The genetic context of these ribozymes allows us to imagine in what biological processes they may be involved (Webb et al. 2009). An HDV-like ribozyme has been identified in the cytoplasmic polyadenylation element binding protein 3 (CPEB3) gene of mammals including humans (Salehi-Ashtiani et al. 2006; Webb and Luptak 2011). This protein binds to the 3' ends of target messengers via cassettes of nucleotides, forming a motif called the cytoplasmic polyadenylation element

(CPE) that allows it to control the polyadenylation step of the messengers which directly influences their ability to be translated (Richter 2007). In the absence of the polyA end, the messenger is either stored in a dormant form – to be translated "later" – or degraded, in P-bodies by the RNA interference mechanism (Ford et al. 2019). In mammals, this gene is involved in long-term memory at the hippocampal brain structure (Vogler et al. 2009; Stephan et al. 2015; Bendixsen et al. 2021). Normally, CPEB3 has a specific post-translational modification, the addition of a SUMO (Small Ubiquitin-like Modifier) domain that renders it soluble and inactive in P-bodies. When the SUMO domain is cleaved following synaptic stimulation, CPEB3 proteins aggregate via their prion domain and interact with actin filaments of the cytoskeleton. They also bind to their target messenger RNAs, which are then polyadenylated and translated. These aggregates are transported to the synapse by the cytoskeleton in order to concentrate the molecules necessary for the memory process. Thus, upon neuronal activation, an immediate response is triggered, saving the step of chromatin activation and protein transport along the dendrites (Drisaldi et al. 2015; Stephan et al. 2015). These mechanisms relying on the relocalization to the synapse of molecules involved in memory processes could explain some aspects of memory according to a molecular model, including the notion of entrainment related to deep and repeated learning that, by stimulating synapses, promote the concentration of spatiotemporal molecules necessary for memory maintenance.

The ribozyme is located in the second intron of the pre-messenger and can therefore in principle be active as soon as the gene is transcribed (Salehi-Ashtiani et al. 2006). One study showed that individuals with a polymorphism of a uridine (U) to cytosine (C) residue at a critical position (nucleotide 36 whose nucleobase pairs with the 5' guanine of the ribozyme) performed worse than the U group in long-term learning of word sets (Vogler et al. 2009) (Figure 7.5). These observations could be related to the fact that the ribozyme with this polymorphism cuts more efficiently than its U36 counterpart. This more efficient cutting would more significantly decrease the amount of messenger RNA and thus the amount of CPEB3 protein.

The team of Andrej Luptak, who co-discovered this ribozyme with Jack Szostak, studied its activity in vitro and in cellulo (Chen et al. 2021a). Their results indicate that under standard conditions, the ribozyme cleavage kinetics are in step with the transcription rate. The ribozyme cleavage rate is thus slow enough for the polymerase to reach the third exon (10 kb to be transcribed at a rate of about ~70 nt/s) and for splicing to allow proper maturation into messenger RNAs of some of the pre-messengers. When ribozyme is blocked by an antisense oligonucleotide (ASO, which hybridizes to the ribozyme by forming a DNA-RNA helix and prevents it from adopting its active structure while leading it to a degradation process by RNase H), the amounts of CPEB3 messengers and proteins increase significantly confirming the basal activity of ribozyme under standard conditions. These effects observed in vitro are also observed in cultures of primary chimpanzee cortical neurons stimulated with potassium chloride (KCl, which causes plasma membrane depolarization) or glutamate (a classical neurotransmitter that targets AMPA (α-amino-3-hydroxy-5-methyl-4-isoazolepropionic acid) receptor). This case of coupling between splicing and cleavage by a ribozyme is reminiscent of the lariat-capping ribozyme that was discussed earlier. Ribozymes are expression regulators in mammals, fungi and also in bacteria if we consider the glmS ribozyme (see Figure 3.10(a)). Considering the evolutionary distance between them, it would be surprising if new examples of this type of regulation were not found in the future.

These experiments therefore show a direct relationship between the level of protein expression and ribozyme activity, and thus between catalytic activity and memory. It is likely that in vivo, a repression mechanism of the activity is set up during a neuronal stimulation. The experiments described point to a mechanism where a pre-messenger sequence upstream of the ribozyme site could act as an ASO. Such mechanisms have already been described for the HDV ribozyme (Diegelman-Parente and Bevilacqua 2002) or for the lariat-capping ribozyme (Nielsen et al. 2009). In the case of HDV, the elongation of the RNA strand during transcription generates a sequence that hybridizes to the 3' strands of P2 and P4, thus preventing the formation of the catalytically competent structure. The LC ribozyme, on the other hand, has its triple junction structure destabilized by a hairpin that competes with the P2 element. Another mechanism could be based on interactions between ribozymes that would lead to the formation of oligomers of pre-messengers, since these are neighboring to each other during synthesis and thus have increased probabilities to interact. The crystallographic structure of the CPEB3 ribozyme shows that ribozyme dimers are formed via the L3 loops, which play a key role in catalytic site formation (Przytula-Mally et al. 2022). Thus, these dimers trap ribozymes in inactive conformations and would promote messenger maturation and thus CPEB3 protein expression. This hypothesis is reinforced by publications of several examples of ribozyme dimers (Suslov et al. 2015; Lunse et al. 2017), although in these cases the dimers promote catalysis. However, given what is known about the subject, it is not possible to decide between these different regulatory mechanisms.

Still other ribozymes have been described relatively recently (Figure 7.6). One has been characterized by screening the human transcriptome using an approach that detects the 5'-OH cleavage product of a hammerhead ribozyme. Interestingly, this second approach revalidated the CPEB3 ribozyme, while identifying a region of a very large (96 kb) regulatory non-coding RNA belonging to an intergenic sequence (vlincRNA: very long intergenic non-coding RNA) that also undergoes a high cleavage rate (Chen et al. 2021b). This novel ribozyme was named Hovlinc because it is located in a human vlincRNA (hominin-vlincRNA-located). Phylogenetic analyses indicate its presence exclusively in placental mammals and not in other vertebrates meaning it is possible to date the appearance of this vlincRNA at about 130 million years. Only the human, chimpanzee and bonobo versions show catalytic activity under the conditions tested by the authors.

In contrast, the gorilla version is not catalytic, indicating that this non-coding RNA was converted to ribozyme about 45 million years ago. The mechanisms of action of lncRNAs are still poorly understood. They are expressed specifically in certain tissues at key stages of cell differentiation and at particular times during development, for example, modifying the local transcriptional state of chromatin (Long et al. 2017; Statello et al. 2021) but not exclusively. They can also recruit regulatory proteins at the level of a gene or a whole chromosome as in silencing the second X chromosome of female mammals, inhibit the binding of transcription factors to their targets, or on the other hand, recruit factors during transcription (Long et al. 2017). Thus, further research is needed to better understand the function of the Hovlinc ribozyme.

It is important to note that in these studies (Salehi-Ashtiani et al. 2006; Hernandez et al. 2020; Chen et al. 2021b), screening of libraries from RNA following various approaches has identified several dozen loci with potential ribozymes. In general, the authors chose to validate the candidate that emerged most widely by characterizing RNA cleavage and its independence from proteins that might act as a nuclease. These studies therefore hold untapped potential as many candidates have not yet been published. This prospect suggests that new ribozymes may yet be discovered in the human genome.

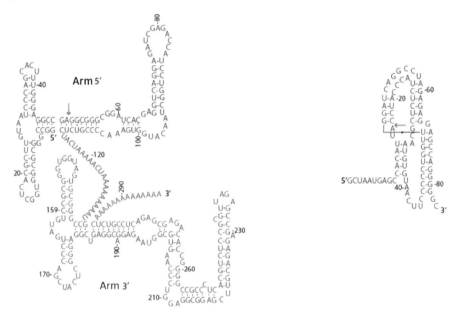

Figure 7.6. *Ribozymes in the human genome*

COMMENTARY ON FIGURE 7.6.– *(a) The Alu ribozyme (ALUr) shown is that of the Y family sequences (Hernandez et al. 2020). The location of the high-throughput sequencing-mapped cut of the 5' fragments of the 5' arm (purple) of the repeat element is indicated by an orange arrow. The Y element is more active than elements in the S and J families. Based on in vitro experiments, the 3' arm also undergoes a cut, but the homology of sequences between the 5' and 3' arms does not allow them to be distinguished and thus mapped accurately. (b) Like the HDV ribozyme, the Hovlinc ribozyme comprises two pseudoknots formed between two loops. This structure is therefore extremely constrained and compact. While ALUr has been identified as an RNA overexpressed in case of thermal shock, the Hovlinc ribozyme has been detected by a high-throughput sequencing approach of an RNA library consisting of fragments resulting from a cutting activity generating fragments with a 5'-OH end typical of endonucleolytic ribozymes. This approach allowed the authors to find the CPEB3 ribozyme and also to identify several sites of interest that will be studied in the future with a high potential to characterize ribozymes not yet described.*

Conclusion

The previous few examples are emblematic of where the field of ribozyme research currently stands. Historical ribozymes (group I and II introns, RNase P, autonucleolytic ribozymes) have provided a better understanding of how RNA chain folding shapes a catalytically active site and have allowed the study of catalytic mechanisms. These studies have also elucidated molecular strategies for activating the reaction, stabilizing the transition states and finally stabilizing the leaving groups. Autonucleolytic ribozymes have been identified in biological systems whose function was to replicate a circular RNA. The advent of high-throughput sequencing methods, by identifying these historical ribozymes in most genomes, raises new questions about their implications in the mechanism of regulating gene expression, subcellular compartmentalization of biological processes in compartments without membranes such as nucleoli or P-bodies. Indeed, the only genomic ribozyme whose function seems to be clearly established today is glmS, which is necessary as a cofactor for the reaction of the product of the enzyme whose expression it controls. However, the development of research projects on ribozymes in basic biology is still rare. Ribozymes are often considered as molecular biology tools and used as nucleases capable of targeting particular sequences by complementarity. In Table 6.1, gene splicing, translation or tRNA maturation are the most visible phenomena orchestrated by ribozymes whose origins date back hundreds of millions of years (Rogers 2019). The fact that these enzymatic activities result from the action of ribonucleoprotein particles has long obscured the catalytic role of RNAs. Indeed, proteins were considered the only biological polymers with catalytic activity.

Over the past two decades, new examples of catalytic RNAs have been regularly identified by deep sequencing techniques (e.g. CPEB3 and Hovlinc ribozymes); however, it remains difficult to identify their precise functions. The human genome contains less than 2% of protein-coding sequences. The rest is RNA, 50% of which corresponds to retrotransposon-like elements! SINEs, which contain epigenetic ribozymes (Alu), alone account for about 11% of the genome and are important in the early stages of development. The CPEB3 gene also contains a ribozyme whose activity is linked to memory phenomena. The presence of a ribozyme in a vlincRNA is also noteworthy (Hovlinc), since these RNAs are involved in the regulation of tissue gene expression and often involved in the mechanisms of carcinogenesis. Genome analyses clearly show that the protein coding portion is relatively congruent, whereas the transcribed portion distinguishes organisms from one another and increases with the complexity of the organism (Taft et al. 2007). Numerous long non-coding RNAs (lncRNAs) are continuously characterized in transcriptomics studies. These RNAs comprise different domains. They can be spliced, capped and polyadenylated like classical messengers. They function by domains that interact with chromatin to conserve it at rest or, on the contrary, to activate it. They can displace factors involved in epigenetic regulation or transcription (Long et al. 2017; Mattick 2018). For example, the intron containing the CPEB3 ribozyme can be considered a lncRNA, as can the Alu elements or even the Hovlinc ribozyme. Some lncRNAs thus exert their function through the action of ribozymes. Given the interest of the scientific community in lncRNAs, it would not be surprising to discover new ribozymes with various catalytic activities, i.e. not only based on transesterification. RNAs and ribozymes are thus bound to be increasingly studied if we wish to accurately capture the fundamental mechanisms of living organisms and thus better understand, better diagnose and better treat diseases that are still incurable today (Mattick 2018).

References

Adams, P.L., Stahley, M.R., Gill, M.L., Kosek, A.B., Wang, J., Strobel, S.A. (2004a). Crystal structure of a group I intron splicing intermediate. *RNA*, 10(12), 1867–1887.

Adams, P.L., Stahley, M.R., Kosek, A.B., Wang, J., Strobel, S.A. (2004b). Crystal structure of a self-splicing group I intron with both exons. *Nature*, 430(6995), 45–50.

Aizawa, Y., Xiang, Q., Lambowitz, A.M., Pyle, A.M. (2003). The pathway for DNA recognition and RNA integration by a group II intron retrotransposon. *Mol. Cell.*, 11(3), 795–805.

Altman, S. (1971). Isolation of tyrosine tRNA precursor molecules. *Nat. New Biol.*, 229(1), 19–21.

Altman, S. (ed.) (1989). Enzymatic cleavage of RNA by RNA. Nobel Lecture, Stockholm.

Altman, S. and Smith, J.D. (1971). Tyrosine tRNA precursor molecule polynucleotide sequence. *Nat. New Biol.*, 233(36), 35–39.

Amunts, A., Brown, A., Bai, X.C., Llacer, J.L., Hussain, T., Emsley, P., Long, F., Murshudov, G., Scheres, S.H., Ramakrishnan, V. (2014). Structure of the yeast mitochondrial large ribosomal subunit. *Science*, 343(6178), 1485–1489.

Andersen, K.L., Beckert, B., Masquida, B., Johansen, S.D., Nielsen, H. (2016). Accumulation of stable full-length circular group I intron RNAs during heat-shock. *Molecules*, 21(11), 1451.

Anderson, M., Schultz, E.P., Martick, M., Scott, W.G. (2013). Active-site monovalent cations revealed in a 1.55-A-resolution hammerhead ribozyme structure. *J. Mol. Biol.*, 425(20), 3790–3798.

Asha, K., Kumar, P., Sanicas, M., Meseko, C.A., Khanna, M., Kumar, B. (2018). Advancements in nucleic acid based therapeutics against respiratory viral infections. *J. Clin. Med.*, 8(1), 6.

Attwater, J., Wochner, A., Holliger, P. (2013). In-ice evolution of RNA polymerase ribozyme activity. *Nat. Chem.*, 5(12), 1011–1018.

Auffinger, P. and Hashem, Y. (2007). Nucleic acid solvation: From outside to insight. *Curr. Opin. Struct. Biol.*, 17(3), 325–333.

Auffinger, P., Grover, N., Westhof, E. (2011). Metal ion binding to RNA. *Met. Ions Life Sci.*, 9, 1–35.

Auffinger, P., D'Ascenzo, L., Ennifar, E. (2016). Sodium and potassium interactions with nucleic acids. *Met. Ions Life Sci.*, 16, 167–201.

Auffinger, P., Ennifar, E., D'Ascenzo, L. (2020). Deflating the RNA Mg(2+) bubble. Stereochemistry to the rescue! *RNA*, 27(3), 243–252.

Baillie, J.K., Barnett, M.W., Upton, K.R., Gerhardt, D.J., Richmond, T.A., De Sapio, F., Brennan, P.M., Rizzu, P., Smith, S., Fell, M. et al. (2011). Somatic retrotransposition alters the genetic landscape of the human brain. *Nature*, 479(7374), 534–537.

Baird, N.J., Zhang, J., Hamma, T., Ferré-D'Amaré, A.R. (2012). YbxF and YlxQ are bacterial homologs of L7Ae and bind K-turns but not K-loops. *RNA*, 18(4), 759–770.

Baldauf, S.L. (2003). The deep roots of eukaryotes. *Science*, 300(5626), 1703–1706.

Balke, D., Wichert, C., Appel, B., Müller, S. (2014). Generation and selection of ribozyme variants with potential application in protein engineering and synthetic biology. *Appl. Microbiol. Biotechnol.*, 98(8), 3389–3399.

Ban, N., Nissen, P., Hansen, J., Moore, P.B., Steitz, T.A. (2000). The complete atomic structure of the large ribosomal subunit at 2.4 A resolution. *Science*, 289(5481), 905–920.

Barrick, J.E., Corbino, K.A., Winkler, W.C., Nahvi, A., Mandal, M., Collins, J., Lee, M., Roth, A., Sudarsan, N., Jona, I. et al. (2004). New RNA motifs suggest an expanded scope for riboswitches in bacterial genetic control. *Proc. Natl. Acad. Sci. USA*, 101(17), 6421–6426.

Beckert, B., Nielsen, H., Einvik, C., Johansen, S.D., Westhof, E., Masquida, B. (2008). Molecular modelling of the GIR1 branching ribozyme gives new insight into evolution of structurally related ribozymes. *EMBO J.*, 27(4), 667–678.

Beese, L.S. and Steitz, T.A. (1991). Structural basis for the 3'-5' exonuclease activity of *Escherichia coli* DNA polymerase I: A two metal ion mechanism. *EMBO J.*, 10(1), 25–33.

Belfort, M. and Lambowitz, A.M. (2019). Group II intron RNPs and reverse transcriptases: From retroelements to research tools. *Cold Spring Harb. Perspect. Biol.*, 11(4).

Bellaousov, S. and Mathews, D.H. (2010). ProbKnot: Fast prediction of RNA secondary structure including pseudoknots. *RNA*, 16(10), 1870–1880.

Bendixsen, D.P., Pollock, T.B., Peri, G., Hayden, E.J. (2021). Experimental resurrection of ancestral mammalian CPEB3 ribozymes reveals deep functional conservation. *Mol. Biol. Evol.*, 38(7), 2843–2853.

Berk, A., Kaiser, C.A., Lodish, H., Amon, A., Ploegh, H., Bretscher, A., Krieger, M., Martin, K.C. (2016). *Molecular Cell Biology*. WH Freeman, New York.

Bershtein, S. and Tawfik, D.S. (2008). Ohno's model revisited: Measuring the frequency of potentially adaptive mutations under various mutational drifts. *Mol. Biol. Evol.*, 25(11), 2311–2318.

Berzal-Herranz, A., Joseph, S., Chowrira, B.M., Butcher, S.E., Burke, J.M. (1993). Essential nucleotide sequences and secondary structure elements of the hairpin ribozyme. *EMBO J.*, 12, 2567–2574.

Birgisdottir, A.B. and Johansen, S. (2005). Site-specific reverse splicing of a HEG-containing group I intron in ribosomal RNA. *Nucleic Acids Res.*, 33(6), 2042–2051.

Birgisdottir, A.B., Nielsen, H., Beckert, B., Masquida, B., Johansen, S.D. (2011). Intermolecular interaction between a branching ribozyme and associated homing endonuclease mRNA. *Biol. Chem.*, 392(6), 491–499.

Boccaletto, P., Stefaniak, F., Ray, A., Cappannini, A., Mukherjee, S., Purta, E., Kurkowska, M., Shirvanizadeh, N., Destefanis, E., Groza, P. et al. (2022). MODOMICS: A database of RNA modification pathways. 2021 update. *Nucleic Acids Res.*, 50(D1), D231–D235.

Bonen, L. (2012). Evolution of mitochondrial introns in plants and photosynthetic microbes. *Adv. Bot. Res.*, 63, 155–186.

Bonen, L. and Vogel, J. (2001). The ins and outs of group II introns. *Trends Genet.*, 17(6), 322–331.

Borah, B., Chen, C.W., Egan, W., Miller, M., Wlodawer, A., Cohen, J.S. (1985). Nuclear magnetic resonance and neutron diffraction studies of the complex of ribonuclease A with uridine vanadate, a transition-state analog. *Biochemistry*, 24(8), 2058–2067.

Brown, R.S., Dewan, J.C., Klug, A. (1985). Crystallographic and biochemical investigation of the lead(II)-catalyzed hydrolysis of yeast phenylalanine tRNA. *Biochemistry*, 24(18), 4785–4801.

Buzayan, J.M., Gerlach, W.L., Bruening, G. (1986). Non-enzymatic cleavage and ligation of RNAs complementary to a plant virus satellite RNA. *Nature*, 323, 349–353.

Canny, M.D., Jucker, F.M., Kellogg, E., Khvorova, A., Jayasena, S.D., Pardi, A. (2004). Fast cleavage kinetics of a natural hammerhead ribozyme. *J. Am. Chem. Soc.*, 126(35), 10848–10849.

Cate, J.H., Gooding, A.R., Podell, E., Zhou, K., Golden, B.L., Kundrot, C.E., Cech, T.R., Doudna, J.A. (1996a). Crystal structure of a group I ribozyme domain: Principles of RNA packing. *Science*, 273, 1678–1684.

Cate, J.H., Gooding, A.R., Podell, E., Zhou, K., Golden, B.L., Szewczak, A.A., Kundrot, C.E., Cech, T.R., Doudna, J.A. (1996b). RNA tertiary structure mediation by adenosine platforms. *Science*, 273, 1696–1699.

Cech, T.R. (ed.) (1989). Self-splicing and enzymatic activity of an intervening sequence RNA from *Tetrahymena*. Nobel Lecture, Stockholm.

Cech, T.R., Zaug, A.J., Grabowski, P.J. (1981). In vitro splicing of the ribosomal RNA precursor of *Tetrahymena*: Involvement of a guanosine nucleotide in the excision of the intervening sequence. *Cell*, 27, 487–496.

Cervera, A. and De La Pena, M. (2014). Eukaryotic penelope-like retroelements encode hammerhead ribozyme motifs. *Mol. Biol. Evol.*, 31(11), 2941–2947.

Cervera, A., Urbina, D., De La Pena, M. (2016). Retrozymes are a unique family of non-autonomous retrotransposons with hammerhead ribozymes that propagate in plants through circular RNAs. *Genome Biol.*, 17(1), 135.

Chang, D.D. and Clayton, D.A. (1989). Mouse RNAase MRP RNA is encoded by a nuclear gene and contains a decamer sequence complementary to a conserved region of mitochondrial RNA substrate. *Cell*, 56(1), 131–139.

Chao, H.W., Tsai, L.Y., Lu, Y.L., Lin, P.Y., Huang, W.H., Chou, H.J., Lu, W.H., Lin, H.C., Lee, P.T., Huang, Y.S. (2013). Deletion of CPEB3 enhances hippocampus-dependent memory via increasing expressions of PSD95 and NMDA receptors. *J. Neurosci.*, 33(43), 17008–17022.

Cheah, M.T., Wachter, A., Sudarsan, N., Breaker, R.R. (2007). Control of alternative RNA splicing and gene expression by eukaryotic riboswitches. *Nature*, 447(7143), 497–500.

Chen, J.H., Yajima, R., Chadalavada, D.M., Chase, E., Bevilacqua, P.C., Golden, B.L. (2010a). A 1.9 A crystal structure of the HDV ribozyme precleavage suggests both Lewis acid and general acid mechanisms contribute to phosphodiester cleavage. *Biochemistry*, 49(31), 6508–6518.

Chen, R., Linnstaedt, S.D., Casey, J.L. (2010b). RNA editing and its control in hepatitis delta virus replication. *Viruses*, 2(1), 131–146.

Chen, C.C., Han, J., Chinn, C.A., Li, X., Nikan, M., Myszka, M., Tong, L., Bredy, T.W., Wood, M.A., Lupták, A. (2021a). The CPEB3 ribozyme modulates hippocampal-dependent memory. *bioRxiv*, 2021.2001.2023.426448.

Chen, Y., Qi, F., Gao, F., Cao, H., Xu, D., Salehi-Ashtiani, K., Kapranov, P. (2021b). Hovlinc is a recently evolved class of ribozyme found in human lncRNA. *Nat. Chem. Biol.*, 17(5), 601–607.

Chinwalla, A.T., Cook, L.L., Delehaunty, K.D., Fewell, G.A., Fulton, L.A., Fulton, R.S., Graves, T.A., Hillier, L.W., Mardis, E.R., McPherson, J.D. et al. (2002). Initial sequencing and comparative analysis of the mouse genome. *Nature*, 420(6915), 520–562.

Cho, I.M., Lai, L.B., Susanti, D., Mukhopadhyay, B., Gopalan, V. (2010). Ribosomal protein L7Ae is a subunit of archaeal RNase P. *Proc. Natl. Acad. Sci. USA*, 107(33), 14573–14578.

Claudet, J., Bopp, L., Cheung, W.W., Devillers, R., Escobar-Briones, E., Haugan, P., Heymans, J.J., Masson-Delmotte, V., Matz-Lück, N., Miloslavich, P. et al. (2020). A roadmap for using the UN decade of ocean science for sustainable development in support of science, policy, and action. *One Earth*, 2(1), 34–42.

Cochrane, J. and Strobel, S.A. (2008). Catalytic strategies of self-cleaving ribozymes. *Acc. Chem. Res.*, 41(8), 1027–1035.

Cochrane, J., Lipchock, S.V., Strobel, S.A. (2007). Structural investigation of the GlmS ribozyme bound to Its catalytic cofactor. *Chem. Biol.*, 14(1), 97–105.

Cochrane, J., Lipchock, S., Smith, K., Strobel, S.A. (2009). Structural and chemical basis for glucosamine-6-phosphate binding and activation of the glmS ribozyme. *Biochemistry*, 48(15), 3239–3246.

Cojocaru, R. and Unrau, P.J. (2021). Processive RNA polymerization and promoter recognition in an RNA World. *Science*, 371(6535), 1225–1232.

Cole, K.H. and Lupták, A. (2019). High-throughput methods in aptamer discovery and analysis. *Methods in Enzymology*, 621(2019), 329–346.

Collins, J.A., Irnov, I., Baker, S., Winkler, W.C. (2007). Mechanism of mRNA destabilization by the glmS ribozyme. *Genes Dev.*, 21(24), 3356–3368.

Correll, C.C., Freeborn, B., Moore, P.B., Steitz, T.A. (1997). Metals, motifs, and recognition in the crystal structure of a 5S rRNA domain. *Cell*, 28, 705–712.

Correll, C.C., Wool, I.G., Munishkin, A. (1999). The two faces of the *Escherichia coli* 23 S rRNA sarcin/ricin domain: The structure at 1.11 A resolution. *J. Mol. Biol.*, 292(2), 275–287.

Costa, M. and Michel, F. (1995). Frequent use of the same tertiary motif by self-folding RNAs. *EMBO J.*, 14(6), 1276–1285.

Costa, M., Walbott, H., Monachello, D., Westhof, E., Michel, F. (2016). Crystal structures of a group II intron lariat primed for reverse splicing. *Science*, 354(6316), aaf9258.

Crary, S.M., Niranjanakumari, S., Fierke, C.A. (1998). The protein component of *Bacillus subtilis* ribonuclease P increases catalytic efficiency by enhancing interactions with the 5' leader sequence of pre-tRNAAsp. *Biochemistry*, 37(26), 9409–9416.

Crick, F. (1966). Codon-anticodon pairing: The wobble hypothesis. *J. Mol. Biol.*, 19, 548–555.

Crick, F. (1968). The origin of the genetic code. *J. Mol. Biol.*, 38, 367–379.

Curtis, E.A. and Bartel, D.P. (2001). The hammerhead cleavage reaction in monovalent cations. *RNA*, 7(4), 546–552.

D'Ascenzo, L. and Auffinger, P. (2016). Anions in nucleic acid crystallography. *Methods Mol. Biol.*, 1320, 337–351.

D'Ascenzo, L., Leonarski, F., Vicens, Q., Auffinger, P. (2017). Revisiting GNRA and UNCG folds: U-turns versus Z-turns in RNA hairpin loops. *RNA*, 23(3), 259–269.

D'Ascenzo, L., Vicens, Q., Auffinger, P. (2018). Identification of receptors for UNCG and GNRA Z-turns and their occurrence in rRNA. *Nucleic Acids Res.*, 46(15), 7989–7997.

Dahm, S.C., Derrick, W.B., Uhlenbeck, O.C. (1993). Evidence for the role of solvated metal hydroxide in the hammerhead cleavage mechanism. *Biochemistry*, 32(48), 13040–13045.

Darr, S.C., Pace, B., Pace, N.R. (1990). Characterization of ribonuclease P from the archaebacterium *Sulfolobus solfataricus*. *J. Biol. Chem.*, 265(22), 12927–12932.

Das, S.R. and Piccirilli, J.A. (2005). General acid catalysis by the hepatitis delta virus ribozyme. *Nat. Chem. Biol.*, 1(1), 45–52.

Dawson, W.K., Fujiwara, K., Kawai, G. (2007). Prediction of RNA pseudoknots using heuristic modeling with mapping and sequential folding. *PLoS One*, 2(9), e905.

De La Pena, M. and Garcia-Robles, I. (2010). Intronic hammerhead ribozymes are ultraconserved in the human genome. *EMBO Rep.*, 11(9), 711–716.

De La Pena, M., Ceprian, R., Casey, J.L., Cervera, A. (2021). Hepatitis delta virus-like circular RNAs from diverse metazoans encode conserved hammerhead ribozymes. *Virus Evol.*, 7(1), veab016.

Deininger, P. (2011). Alu elements: Know the SINEs. *Genome Biol.*, 12(12), 236.

Demeshkina, N., Jenner, L., Westhof, E., Yusupov, M., Yusupova, G. (2012). A new understanding of the decoding principle on the ribosome. *Nature*, 484(7393), 256–259.

Demeshkina, N., Jenner, L., Westhof, E., Yusupov, M., Yusupova, G. (2013). New structural insights into the decoding mechanism: Translation infidelity via a G.U pair with Watson-Crick geometry. *FEBS Lett.*, 587(13), 1848–1857.

Diegelman-Parente, A. and Bevilacqua, P.C. (2002). A mechanistic framework for co-transcriptional folding of the HDV genomic ribozyme in the presence of downstream sequence. *J. Mol. Biol.*, 324(1), 1–16.

Doherty, E.A., Batey, R.T., Masquida, B., Doudna, J.A. (2001). A universal mode of helix packing in RNA. *Nat. Struct. Biol.*, 8(4), 339–343.

Drisaldi, B., Colnaghi, L., Fioriti, L., Rao, N., Myers, C., Snyder, A.M., Metzger, D.J., Tarasoff, J., Konstantinov, E., Fraser, P.E. et al. (2015). SUMOylation is an inhibitory constraint that regulates the prion-like aggregation and activity of CPEB3. *Cell Rep.*, 11(11), 1694–1702.

Duan, J., Wang, X., Kizer, M.E. (2020). Biotechnological and therapeutic applications of natural nucleic acid structural motifs. *Top. Curr. Chem.*, 378(2), 26.

Dunham, C.M., Murray, J.B., Scott, W.G. (2003). A helical twist-induced conformational switch activates cleavage in the hammerhead ribozyme. *J. Mol. Biol.*, 332(2), 327–336.

Dunin-Horkawicz, S., Czerwoniec, A., Gajda, M.J., Feder, M., Grosjean, H., Bujnicki, J.M. (2006). MODOMICS: A database of RNA modification pathways. *Nucleic Acids Res.*, 34(Database issue), D145–149.

Eddy, S.R. (2004). What is a hidden Markov model? *Nat. Biotech.*, 22(10), 1315.

Ellington, A.D. and Szostak, J.W. (1990). In vitro selection of RNA molecules that binds specific ligands. *Nature*, 346, 818–822.

Emilsson, G.M., Nakamura, S., Roth, A., Breaker, R.R. (2003). Ribozyme speed limits. *RNA*, 9(8), 907–918.

Evgen'ev, M.B. (2013). What happens when Penelope comes? An unusual retroelement invades a host species genome exploring different strategies. *Mob. Genet. Elements*, 3(2), e24542.

Evgen'ev, M.B. and Arkhipova, I.R. (2005). Penelope-like elements – A new class of retroelements: Distribution, function and possible evolutionary significance. *Cytogenet. Genome Res.*, 110(1/4), 510–521.

Evgen'ev, M.B., Zelentsova, H., Shostak, N., Kozitsina, M., Barskyi, V., Lankenau, D.H., Corces, V.G. (1997). Penelope, a new family of transposable elements and its possible role in hybrid dysgenesis in *Drosophila virilis*. *Proc. Natl. Acad. Sci. USA*, 94(1), 196–201.

Famulok, M. and Mayer, G. (2014). Aptamers and SELEX in Chemistry & Biology. *Chem. Biol.*, 21(9), 1055–1058.

Fedor, M.J. and Uhlenbeck, O.C. (1990). Substrate sequence effects on "hammerhead" RNA catalytic efficiency. *Proc. Natl. Acad. Sci. USA*, 87(5), 1668–1672.

Felletti, M. and Hartig, J.S. (2017). Ligand-dependent ribozymes. *WIREs RNA*, 8(2), e1395.

Ferbeyre, G., Bourdeau, V., Pageau, M., Miramontes, P., Cedergren, R. (2000). Distribution of hammerhead and hammerhead-like RNA motifs through the GenBank. *Genome Res.*, 10(7), 1011–1019.

Ferguson, L.R. and Denny, W.A. (1991). The genetic toxicology of acridines. *Mutat. Res./Rev. Genet. Toxicol.*, 258(2), 123–160.

Ferré-D'Amaré, A.R. (2010). Use of the spliceosomal protein U1A to facilitate crystallization and structure determination of complex RNAs. *Methods*, 52(2), 159–167.

Ferré-D'Amaré, A.R. and Doudna, J.A. (2000). Crystallization and structure determination of a hepatitis delta virus ribozyme: Use of the RNA-binding protein U1A as a crystallization module. *J. Mol. Biol.*, 295(3), 541–556.

Ferré-D'Amaré, A.R., Zhou, K.H., Doudna, J.A. (1998). Crystal structure of a hepatitis delta virus ribozyme. *Nature*, 395(6702), 567–574.

Ferron, F., Sama, B., Decroly, E., Canard, B. (2021). The enzymes for genome size increase and maintenance of large (+)RNA viruses. *Trends Biochem. Sci.*, 46(11), 866–877.

Fica, S.M., Oubridge, C., Galej, W.P., Wilkinson, M.E., Bai, X.C., Newman, A.J., Nagai, K. (2017). Structure of a spliceosome remodelled for exon ligation. *Nature*, 542(7641), 377–380.

Fire, A. (2007). Gene silencing by double-stranded RNA. *Cell Death Differ.*, 14(12), 1998–2012.

Fire, A., Xu, S., Montgomery, M.K., Kostas, S.A., Driver, S.E., Mello, C.C. (1998). Potent and specific genetic interference by double-stranded RNA in *Caenorhabditis elegans*. *Nature*, 391(6669), 806–811.

Flores, R., Gas, M.-E., Molina-Serrano, D., Nohales, M.-Á., Carbonell, A., Gago, S., De La Peña, M., Daròs, J.-A. (2009). Viroid replication: Rolling-circles, enzymes and ribozymes. *Viruses*, 1(2), 317–334.

Flores, R., Grubb, D., Elleuch, A., Nohales, M.A., Delgado, S., Gago, S. (2011). Rolling-circle replication of viroids, viroid-like satellite RNAs and hepatitis delta virus: Variations on a theme. *RNA Biol.*, 8(2), 200–206.

Ford, L., Ling, E., Kandel, E.R., Fioriti, L. (2019). CPEB3 inhibits translation of mRNA targets by localizing them to P bodies. *Proc. Natl. Acad. Sci. USA*, 116(36), 18078–18087.

Forster, A.C., Jeffries, A.C., Sheldon, C.C., Symons, R.H. (1987). Structural and ionic requirements for self-cleavage of virusoid RNAs and trans self-cleavage of viroid RNA. *Cold Spring Harb. Symp. Quant. Biol.*, 52, 249–259.

Frauendorf, C. and Jaschke, A. (1998). Catalysis of organic reactions by RNA. *Angew. Chem. Int. Ed.*, 37(10), 1378–1381.

Freemont, P.S., Friedman, J.M., Beese, L.S., Sanderson, M.R., Steitz, T.A. (1988). Cocrystal structure of an editing complex of Klenow fragment with DNA. *Proc. Natl. Acad. Sci. USA*, 85(23), 8924–8928.

Fujii, K., Susanto, T.T., Saurabh, S., Barna, M. (2018). Decoding the function of expansion segments in ribosomes. *Mol. Cell.*, 72(6), 1013–1020, e1016.

Galej, W.P., Toor, N., Newman, A.J., Nagai, K. (2018). Molecular mechanism and evolution of nuclear pre-mRNA and group II intron splicing: Insights from cryo-electron microscopy structures. *Chem. Rev.*, 118(8), 4156–4176.

Gardner, P.P., Daub, J., Tate, J., Moore, B.L., Osuch, I.H., Griffiths-Jones, S., Finn, R.D., Nawrocki, E.P., Kolbe, D.L., Eddy, S.R. et al. (2011). Rfam: Wikipedia, clans and the "decimal" release. *Nucleic Acids Res.*, 39(Database issue), D141–145.

Gesteland, R.F. and Atkins, J.F. (1993). *The RNA World: The Nature of Modern RNA Suggests a Prebiotic RNA World*. Cold Spring Harbor Laboratory Press, New York.

Gesteland, R.F., Cech, T., Atkins, J.F. (1999). *The RNA World: The Nature of Modern RNA Suggests a Prebiotic RNA*. Cold Spring Harbor Laboratory Press, New York.

Gesteland, R.F., Cech, T., Atkins, J.F. (2006). *The RNA World: The Nature of Modern RNA Suggests a Prebiotic RNA World.* Cold Spring Harbor Laboratory Press, New York.

Giege, R., Juhling, F., Putz, J., Stadler, P., Sauter, C., Florentz, C. (2012). Structure of transfer RNAs: Similarity and variability. *Wiley Interdiscip. Rev. RNA*, 3(1), 37–61.

Gilbert, W. (1986). Origin of life: The RNA world. *Nature*, 319(6055), 618.

Giuliodori, A.M., Di Pietro, F., Marzi, S., Masquida, B., Wagner, R., Romby, P., Gualerzi, C.O., Pon, C.L. (2010). The cspA mRNA is a thermosensor that modulates translation of the cold-shock protein CspA. *Mol. Cell.*, 37(1), 21–33.

Gladyshev, E.A. and Arkhipova, I.R. (2007). Telomere-associated endonuclease-deficient Penelope-like retroelements in diverse eukaryotes. *Proc. Natl. Acad. Sci. USA*, 104(22), 9352–9357.

Goddard, M.R. and Burt, A. (1999). Recurrent invasion and extinction of a selfish gene. *Proc. Natl. Acad. Sci. USA*, 96(24), 13880–13885.

Golden, B.L., Kim, H., Chase, E. (2005). Crystal structure of a phage Twort group I ribozyme-product complex. *Nat. Struct. Mol. Biol.*, 12(1), 82–89.

Gong, B., Klein, D.J., Ferré-D'Amaré, A.R., Carey, P.R. (2011). The glmS ribozyme tunes the catalytically critical pK(a) of its coenzyme glucosamine-6-phosphate. *J. Am. Chem. Soc.*, 133(36), 14188–14191.

Gottesman, S. (2005). Micros for microbes: Non-coding regulatory RNAs in bacteria. *Trends Genet.*, 21(7), 399.

Greber, B.J., Boehringer, D., Leibundgut, M., Bieri, P., Leitner, A., Schmitz, N., Aebersold, R., Ban, N. (2014). The complete structure of the large subunit of the mammalian mitochondrial ribosome. *Nature*, 515(7526), 283–286.

Greber, B.J., Bieri, P., Leibundgut, M., Leitner, A., Aebersold, R., Boehringer, D., Ban, N. (2015). The complete structure of the 55S mammalian mitochondrial ribosome. *Science*, 348(6232), 303–308.

Guerrier-Takada, C. and Altman, S. (1984). Catalytic activity of an RNA molecule prepared by transcription in vitro. *Science*, 223(4633), 285–286.

Guerrier-Takada, C. and Altman, S. (1986). M1 RNA with large terminal deletions retains its catalytic activity. *Cell*, 45(2), 177–183.

Guerrier-Takada, C., Gardiner, K., Marsh, T., Pace, N., Altman, S. (1983). The RNA moiety of ribonuclease P is the catalytic subunit of the enzyme. *Cell*, 35, 849–857.

Guo, F., Gooding, A.R., Cech, T.R. (2004). Structure of the *Tetrahymena* ribozyme; base triple sandwich and metal ion at the active site. *Mol. Cell.*, 16(3), 351–362.

Haas, E.S., Armbruster, D.W., Vucson, B.M., Daniels, C.J., Brown, J.W. (1996). Comparative analysis of ribonuclease P RNA structure in Archaea. *Nucleic Acids Res.*, 24(7), 1252–1259.

Hafez, M. and Hausner, G. (2012). Homing endonucleases: DNA scissors on a mission. *Genome*, 55(8), 553–569.

Hajdin, C.E., Bellaousov, S., Huggins, W., Leonard, C.W., Mathews, D.H., Weeks, K.M. (2013). Accurate SHAPE-directed RNA secondary structure modeling, including pseudoknots. *Proc. Natl. Acad. Sci. USA*, 110(14), 5498–5503.

Hammann, C., Luptak, A., Perreault, J., De La Pena, M. (2012). The ubiquitous hammerhead ribozyme. *RNA*, 18(5), 871–885.

Hang, J., Wan, R., Yan, C., Shi, Y. (2015). Structural basis of pre-mRNA splicing. *Science*, 349(6253), 1191–1198.

Hartmann, T.B., Thiel, D., Dummer, R., Schadendorf, D., Eichmüller, S. (2004). SEREX identification of new tumour-associated antigens in cutaneous T-cell lymphoma. *Br. J. Dermatol.*, 150(2), 252–258.

Haugen, P. and Bhattacharya, D. (2004). The spread of LAGLIDADG homing endonuclease genes in rDNA. *Nucleic Acids Res.*, 32(6), 2049–2057.

Haugen, P., Reeb, V., Lutzoni, F., Bhattacharya, D. (2004). The evolution of homing endonuclease genes and group I introns in nuclear rDNA. *Mol. Biol. Evol.*, 21(1), 129–140.

Haugen, P., Simon, D.M., Bhattacharya, D. (2005a). The natural history of group I introns. *Trends Genet.*, 21(2), 111–119.

Haugen, P., Wikmark, O.G., Vader, A., Coucheron, D.H., Sjottem, E., Johansen, S.D. (2005b). The recent transfer of a homing endonuclease gene. *Nucleic Acids Res.*, 33(8), 2734–2741.

Hernandez, A.J., Zovoilis, A., Cifuentes-Rojas, C., Han, L., Bujisic, B., Lee, J.T. (2020). B2 and ALU retrotransposons are self-cleaving ribozymes whose activity is enhanced by EZH2. *Proc. Natl. Acad. Sci. USA*, 117(1), 415–425.

Hofacker, I.L., Fekete, M., Stadler, P.F. (2002). Secondary structure prediction for aligned RNA sequences. *J. Mol. Biol.*, 319(5), 1059–1066.

Hollenstein, M. (2019). Nucleic acid enzymes based on functionalized nucleosides. *Curr. Opin. Chem. Biol.*, 52, 93–101.

Holzmann, J., Frank, P., Loffler, E., Bennett, K.L., Gerner, C., Rossmanith, W. (2008). RNase P without RNA: Identification and functional reconstitution of the human mitochondrial tRNA processing enzyme. *Cell*, 135(3), 462–474.

Homer (1992). *The Iliad*. Penguin, London.

Homer (2003). *The Odyssey*. Penguin, London.

Hoogsteen, K. (1963). The crystal and molecular structure of a hydrogen-bonded complex. Between 1-methylthymine and 9-methyladenine. *Acta Crystallogr.*, 16, 907–916.

Horning, D.P. and Joyce, G.F. (2016). Amplification of RNA by an RNA polymerase ribozyme. *Proc. Natl. Acad. Sci. USA*, 113(35), 9786–9791.

Hug, L.A., Baker, B.J., Anantharaman, K., Brown, C.T., Probst, A.J., Castelle, C.J., Butterfield, C.N., Hernsdorf, A.W., Amano, Y., Ise, K. et al. (2016). A new view of the tree of life. *Nat. Microbiol.*, 1, 16048.

Hutchins, C.J., Rathjen, P.D., Forster, A.C., Symons, R.H. (1986). Self-cleavage of plus and minus RNA transcripts of avocado sunblotch viroid. *Nucleic Acids Res.*, 14(9), 3629–3640.

Jabbari, H., Wark, I., Montemagno, C. (2018). RNA secondary structure prediction with pseudoknots: Contribution of algorithm versus energy model. *PLoS One*, 13(4), e0194583.

Jabri, E. and Cech, T.R. (1998). In vitro selection of the Naegleria GIR1 ribozyme identifies three base changes that dramatically improve activity. *RNA*, 4(12), 1481–1492.

Jaeger, L. (1993). Les introns auto-catalytiques de groupe I comme modèle d'étude du repliement des acides ribonucléiques. PhD Thesis, Université Louis Pasteur, Strasbourg.

Jimenez, R.M., Polanco, J.A., Luptak, A. (2015). Chemistry and biology of self-cleaving ribozymes. *Trends Biochem. Sci.*, 40(11), 648–661.

Johansen, S. and Haugen, P. (2001). A new nomenclature of group I introns in ribosomal DNA. *RNA*, 7(7), 935–936.

Johansen, S. and Vogt, V.M. (1994). An intron in the nuclear ribosomal DNA of *Didymium iridis* codes for a group I ribozyme and a novel ribozyme that cooperate in self-splicing. *Cell*, 76(4), 725–734.

Johansen, S., Johansen, T., Haugli, F. (1992). Extrachromosomal ribosomal DNA of *Didymium iridis*: Sequence analysis of the large subunit ribosomal RNA gene and sub-telomeric region. *Curr. Genet.*, 22(4), 305–312.

Johansen, S., Einvik, C., Nielsen, H. (2002). DiGIR1 and NaGIR1: Naturally occurring group I-like ribozymes with unique core organization and evolved biological role. *Biochimie*, 84(9), 905–912.

Jones, C.P. and Ferré-D'Amaré, A.R. (2017). Long-range interactions in riboswitch control of gene expression. *Annu. Rev. Biophys.*, 46(1), 455–481.

Jossinet, F., Ludwig, T.E., Westhof, E. (2010). Assemble: An interactive graphical tool to analyze and build RNA architectures at the 2D and 3D levels. *Bioinformatics*, 26(16), 2057–2059.

Kalvari, I., Nawrocki, E.P., Ontiveros-Palacios, N., Argasinska, J., Lamkiewicz, K., Marz, M., Griffiths-Jones, S., Toffano-Nioche, C., Gautheret, D., Weinberg, Z. et al. (2021). Rfam 14: Expanded coverage of metagenomic, viral and microRNA families. *Nucleic Acids Res.*, 49(D1), D192–D200.

Keiper, S., Bebenroth, D., Seelig, B., Westhof, E., Jaschke, A. (2004). Architecture of a Diels-Alderase ribozyme with a preformed catalytic pocket. *Chem. Biol.*, 11(9), 1217–1227.

Kew, M.C., Dusheiko, G.M., Hadziyannis, S.J., Patterson, A. (1984). Does delta infection play a part in the pathogenesis of hepatitis B virus related hepatocellular carcinoma? *Br. Med. J. (Clin. Res. Ed.)*, 288(6432), 1727.

Khvorova, A., Lescoute, A., Westhof, E., Jayasena, S.D. (2003). Sequence elements outside the hammerhead ribozyme catalytic core enable intracellular activity. *Nat. Struct. Biol.*, 10(9), 708–712.

Kimura, M. and Ohta, T. (1974). On some principles governing molecular evolution. *Proc. Natl. Acad. Sci. USA*, 71(7), 2848–2852.

Klein, D.J. and Ferré-D'Amaré, A.R. (2006). Structural basis of glmS ribozyme activation by glucosamine-6-phosphate. *Science*, 313(5794), 1752–1756.

Klein, D.J., Schmeing, T.M., Moore, P.B., Steitz, T.A. (2001). The kink-turn: A new RNA secondary structure motif. *EMBO J.*, 20(15), 4214–4221.

Klein, D.J., Been, M.D., Ferré-D'Amaré, A.R. (2007). Essential role of an active-site guanine in glmS ribozyme catalysis. *J. Am. Chem. Soc.*, 129(48), 14858–14859.

Koculi, E., Cho, S.S., Desai, R., Thirumalai, D., Woodson, S.A. (2012). Folding path of P5abc RNA involves direct coupling of secondary and tertiary structures. *Nucleic Acids Res.*, 40(16), 8011–8020.

Koski, R.A., Bothwell, A.L., Altman, S. (1976). Identification of a ribonuclease P-like activity from human KB cells. *Cell*, 9(1), 101–116.

Krasilnikov, A.S., Yang, X., Pan, T., Mondragon, A. (2003). Crystal structure of the specificity domain of ribonuclease P. *Nature*, 421(6924), 760–764.

Krasilnikov, A.S., Xiao, Y., Pan, T., Mondragon, A. (2004). Basis for structural diversity in homologous RNAs. *Science*, 306(5693), 104–107.

Krogh, N., Pietschmann, M., Schmid, M., Jensen, T.H., Nielsen, H. (2017). Lariat capping as a tool to manipulate the 5' end of individual yeast mRNA species in vivo. *RNA*, 23(5), 683–695.

Kuimelis, R.G. and McLaughlin, L.W. (1996). Ribozyme-mediated cleavage of a substrate analogue containing an internucleotide-bridging 5'-Phosphorothioate: Evidence for the single metal model. *Biochemistry*, 35, 5308–5317.

Kuo, M.Y., Sharmeen, L., Dinter-Gottlieb, G., Taylor, J. (1988). Characterization of self-cleaving RNA sequences on the genome and antigenome of human hepatitis delta virus. *J. Virol.*, 62(12), 4439–4444.

Lambowitz, A.M. and Belfort, M. (1993). Introns as mobile genetic elements. *Annu. Rev. Biochem.*, 62, 587–622.

Lambowitz, A.M. and Belfort, M. (2015). Mobile bacterial group II introns at the crux of eukaryotic evolution. *Microbiol. Spectr.*, 3(1), MDNA3-0050-2014.

Lambowitz, A.M. and Zimmerly, S. (2004). Mobile group II introns. *Annu. Rev. Genet.*, 38, 1–35.

Lan, P., Tan, M., Zhang, Y., Niu, S., Chen, J., Shi, S., Qiu, S., Wang, X., Peng, X., Cai, G. et al. (2018). Structural insight into precursor tRNA processing by yeast ribonuclease P. *Science*, 362(6415), eaat6678.

Lau, M.W. and Ferré-D'Amaré, A.R. (2016). In vitro evolution of coenzyme-independent variants from the glmS ribozyme structural scaffold. *Methods*, 106, 76–81.

Lawrence, M.S. and Bartel, D.P. (2005). New ligase-derived RNA polymerase ribozymes. *RNA*, 11(8), 1173–1180.

Leclerc, F. and Karplus, M. (2006). Two-metal-ion mechanism for hammerhead-ribozyme catalysis. *J. Phys. Chem. B*, 110(7), 3395–3409.

Leclerc, F., Zaccai, G., Vergne, J., Rihova, M., Martle, A., Maurel, M.-C. (2016). Self-assembly controls self-cleavage of HHR from ASBVd(-): A combined SANS and modeling study. *Sci. Rep.*, 6(30287), doi:10.1038/srep30287.

Lee, E.R., Baker, J.L., Weinberg, Z., Sudarsan, N., Breaker, R.R. (2010). An allosteric self-splicing ribozyme triggered by a bacterial second messenger. *Science*, 329(5993), 845–848.

Lee, C.H., Han, S.R., Lee, S.W. (2018). Group I intron-based therapeutics through trans-splicing reaction. *Prog. Mol. Biol. Transl. Sci.*, 159, 79–100.

Lehnert, V., Jaeger, L., Michel, F., Westhof, E. (1996). New loop–loop tertiary interactions in self-splicing introns of subgroup IC and ID: A complete 3D model of the *Tetrahymena thermophila* ribozyme. *Chem. Biol.*, 3(12), 993–1009.

Leonarski, F., D'Ascenzo, L., Auffinger, P. (2017). Mg2+ ions: Do they bind to nucleobase nitrogens? *Nucleic Acids Res.*, 45(2), 987–1004.

Leonarski, F., D'Ascenzo, L., Auffinger, P. (2019). Nucleobase carbonyl groups are poor Mg(2+) inner-sphere binders but excellent monovalent ion binders – A critical PDB survey. *RNA*, 25(2), 173–192.

Leontis, N.B. and Westhof, E. (2001). Geometric nomenclature and classification of RNA base pairs. *RNA*, 7(4), 499–512.

Lescoute, A. and Westhof, E. (2006). The A-minor motifs in the decoding recognition process. *Biochimie*, 88(8), 993–999.

Levinthal, C. (1968). Are there pathways for protein folding? *J. Chim. Phys.*, 65, 44–47.

Li, Y. and Breaker, R.R. (1999). Kinetics of RNA degradation by specific base catalysis of transesterification involving the 2'-hydroxyl group. *J. Am. Chem. Soc.*, 121(23), 5364–5372.

Li, W.-H., Gu, Z., Wang, H., Nekrutenko, A. (2001). Evolutionary analyses of the human genome. *Nature*, 409(6822), 847–849.

Liberman, J.A., Suddala, K.C., Aytenfisu, A., Chan, D., Belashov, I.A., Salim, M., Mathews, D.H., Spitale, R.C., Walter, N.G., Wedekind, J.E. (2015). Structural analysis of a class III preQ1 riboswitch reveals an aptamer distant from a ribosome-binding site regulated by fast dynamics. *Proc. Natl. Acad. Sci. USA*, 112(27), E3485–3494.

Lilley, D.M. (2017). How RNA acts as a nuclease: Some mechanistic comparisons in the nucleolytic ribozymes. *Biochem. Soc. Trans.*, 45(3), 683–691.

Lilley, D.M. and Norman, D.G. (1999). The Holliday junction is finally seen with crystal clarity. *Nat. Struct. Biol.*, 6(10), 897–899.

Liu, Y., Wilson, T.J., McPhee, S.A., Lilley, D.M. (2014). Crystal structure and mechanistic investigation of the twister ribozyme. *Nat. Chem. Biol.*, 10(9), 739–744.

Liu, Y., Wilson, T.J., Lilley, D.M. (2017). The structure of a nucleolytic ribozyme that employs a catalytic metal ion. *Nat. Chem. Biol.*, 13(5), 508–513.

Long, Y., Wang, X., Youmans, D.T., Cech, T.R. (2017). How do lncRNAs regulate transcription? *Sci. Adv.*, 3(9), eaao2110.

Lorenz, R., Bernhart, S.H., Höner Zu Siederdissen, C., Tafer, H., Flamm, C., Stadler, P.F., Hofacker, I.L. (2011). ViennaRNA Package 2.0. *Algorithms Mol. Biol.*, 6(1), 26.

Lott, W.B., Pontius, B.W., Von Hippel, P.H. (1998). A two-metal ion mechanism operates in the hammerhead ribozyme-mediated cleavage of an RNA substrate. *Proc. Natl. Acad. Sci. USA*, 95(2), 542.

Lunse, C.E., Weinberg, Z., Breaker, R.R. (2017). Numerous small hammerhead ribozyme variants associated with Penelope-like retrotransposons cleave RNA as dimers. *RNA Biol.*, 14(11), 1499–1507.

Luptak, A. (2016). In vitro selection and evolution. *Methods*, 106, 1–2.

Luptak, A., Ferré-D'Amaré, A.R., Zhou, K., Zilm, K.W., Doudna, J.A. (2001). Direct pK(a) measurement of the active-site cytosine in a genomic hepatitis delta virus ribozyme. *J. Am. Chem. Soc.*, 123(35), 8447–8452.

Mandal, M., Boese, B., Barrick, J.E., Winkler, W.C., Breaker, R.R. (2003). Riboswitches control fundamental biochemical pathways in *Bacillus subtilis* and other bacteria. *Cell*, 113(5), 577–586.

Margueron, R. and Reinberg, D. (2011). The Polycomb complex PRC2 and its mark in life. *Nature*, 469(7330), 343–349.

Martick, M. and Scott, W.G. (2006). Tertiary contacts distant from the active site prime a ribozyme for catalysis. *Cell*, 126(2), 309–320.

Martick, M., Horan, L.H., Noller, H.F., Scott, W.G. (2008a). A discontinuous hammerhead ribozyme embedded in a mammalian messenger RNA. *Nature*, 454(7206), 899–902.

Martick, M., Lee, T.S., York, D.M., Scott, W.G. (2008b). Solvent structure and hammerhead ribozyme catalysis. *Chem. Biol.*, 15(4), 332–342.

Masquida, B., Jossinet, F., Westhof, E. (2010). Over a decade of ribonuclease P modelling. In *Protein Reviews*, Liu, F. and Altman, S. (eds). Springer, New York.

Massire, C., Jaeger, L., Westhof, E. (1997). Phylogenetic evidence for a new tertiary interaction in bacterial RNase P RNA. *RNA*, 3, 553–556.

Massire, C., Jaeger, L., Westhof, E. (1998). Derivation of the three-dimensional architecture of bacterial ribonuclease P RNAs from comparative sequence analysis. *J. Mol. Biol.*, 279(4), 773–793.

Mattick, J.S. (2018). The state of long non-coding RNA biology. *Noncoding RNA*, 4(3), 17.

Mattick, J.S. and Makunin, I.V. (2005). Small regulatory RNAs in mammals. *Hum. Mol. Genet.*, 14(1), R121–132.

Maurel, M.-C., Leclerc, F., Vergne, J., Zaccai, G. (2019). RNA back and forth: Looking through ribozyme and viroid motifs. *Viruses*, 11(283), v11030283.

McClain, W.H. (1977). Seven terminal steps in a biosynthetic pathway leading from DNA to transfer RNA. *Acc. Chem. Res.*, 10(11), 418–425.

McKinney, S.A., Tan, E., Wilson, T.J., Nahas, M.K., Declais, A.C., Clegg, R.M., Lilley, D.M., Ha, T. (2004). Single-molecule studies of DNA and RNA four-way junctions. *Biochem. Soc. Trans.*, 32(Pt 1), 41–45.

Meyer, M. and Masquida, B. (2014). Cis-acting 5' hammerhead ribozyme optimization for in vitro transcription of highly structured RNAs. *Methods Mol. Biol.*, 1086, 21–40.

Meyer, M., Westhof, E., Masquida, B. (2012). A structural module in RNase P expands the variety of RNA kinks. *RNA Biol.*, 9(3), 254–260.

Meyer, M., Nielsen, H., Olieric, V., Roblin, P., Johansen, S.D., Westhof, E., Masquida, B. (2014). Speciation of a group I intron into a lariat capping ribozyme. *Proc. Natl. Acad. Sci. USA*, 111(21), 7659–7664.

Meyer, M., Walbott, H., Olieric, V., Kondo, J., Costa, M., Masquida, B. (2019). Conformational adaptation of UNCG loops upon crowding. *RNA*, 25, 1522–1531.

Michel, F. and Dujon, B. (1983). Conservation of RNA secondary structure in two intron families including mitochondrial-, chloroplast-, and nuclear-encoded members. *EMBO J.*, 2, 33–34.

Michel, F. and Westhof, E. (1990). Modelling of the three-dimensional architecture of group-I catalytic introns based on comparative sequence analysis. *J. Mol. Biol.*, 216, 585–610.

Michel, F., Jaquier, A., Dujon, B. (1982). Comparison of fungal mitochondrial introns reveals extensive homologies in RNA secondary structure. *Biochimie*, 64, 867–881.

Michel, F., Ellington, A., Couture, S., Szostak, J.W. (1990). Phylogenetic and genetic evidence for base-triples in the catalytic domain of group I introns. *Nature*, 347, 578–580.

Mir, A. and Golden, B.L. (2016). Two active site divalent ions in the crystal structure of the hammerhead ribozyme bound to a transition state analogue. *Biochemistry*, 55(4), 633–636.

Mir, A., Chen, J., Robinson, K., Lendy, E., Goodman, J., Neau, D., Golden, B.L. (2015). Two divalent metal ions and conformational changes play roles in the hammerhead ribozyme cleavage reaction. *Biochemistry*, 54(41), 6369–6381.

Montange, R.K. and Batey, R.T. (2008). Riboswitches: Emerging themes in RNA structure and function. *Annu. Rev. Biophys.*, 37, 117–133.

Moore, S.M., Skowronska-Krawczyk, D., Chao, D.L. (eds) (2019). Emerging concepts for RNA therapeutics for inherited retinal disease. In *Retinal Degenerative Diseases*, Bowes Rickman, C., Grimm, C., Anderson, R.E., Ash, J.D., LaVail, M.M., Hollyfield, J.G. (eds). Springer International Publishing, Cham.

Mullis, K.B. (1990a). Target amplification for DNA analysis by the polymerase chain reaction. *Ann. Biol. Clin. (Paris)*, 48(8), 579–582.

Mullis, K.B. (1990b). The unusual origin of the polymerase chain reaction. *Sci. Am.*, 262(4), 56–61, 64–55.

Mullis, K.B. (ed.) (1993). The polymerase chain reaction. Nobel Lecture, Stockholm.

Murakami, H., Ohta, A., Ashigai, H., Suga, H. (2006a). A highly flexible tRNA acylation method for non-natural polypeptide synthesis. *Nat. Methods*, 3(5), 357–359.

Murakami, H., Ohta, A., Goto, Y., Sako, Y., Suga, H. (2006b). Flexizyme as a versatile tRNA acylation catalyst and the application for translation. *Nucleic Acids Symp. Ser. (Oxf.)*, 50, 35–36.

Murchie, A.I.H., Thomson, J.B., Walter, F., Lilley, D.M. (1998). Folding of the hairpin in its natural conformation achieves close physical proximity of the loops. *Mol. Cell.*, 1(6), 873–881.

Murray, J.B., Seyhan, A.A., Walter, N.G., Burke, J.M., Scott, W.G. (1998a). The hammerhead, hairpin and VS ribozymes are catalytically proficient in monovalent cations alone. *Chem. Biol.*, 5(10), 587–595.

Murray, J.B., Terwey, D.P., Maloney, L., Karpeisky, A., Usman, N., Beigelman, L., Scott, W.G. (1998b). The structural basis of hammerhead ribozyme self-cleavage. *Cell*, 92(5), 665–673.

Murray, J.B., Szoke, H., Szoke, A., Scott, W.G. (2000). Capture and visualization of a catalytic RNA enzyme-product complex using crystal lattice trapping and X-ray holographic reconstruction. *Mol. Cell.*, 5(2), 279–287.

Nakano, S., Chadalavada, D.M., Bevilacqua, P.C. (2000). General acid-base catalysis in the mechanism of a hepatitis delta virus ribozyme. *Science*, 287(5457), 1493–1497.

Nawrocki, E.P. and Eddy, S.R. (2013). Infernal 1.1: 100-fold faster RNA homology searches. *Bioinformatics*, 29(22), 2933–2935.

Nechooshtan, G., Elgrably-Weiss, M., Sheaffer, A., Westhof, E., Altuvia, S. (2009). A pH-responsive riboregulator. *Genes Dev.*, 23(22), 2650–2662.

Nelson, J.A. and Uhlenbeck, O.C. (2006). When to believe what you see. *Mol. Cell*, 23(4), 447–450.

Nielsen, H. (2011). Working with RNA. *Methods Mol. Biol.*, 703, 15–28.

Nielsen, H., Westhof, E., Johansen, S. (2005). An mRNA is capped by a 2', 5' lariat catalyzed by a group I-like ribozyme. *Science*, 309(5740), 1584–1587.

Nielsen, H., Einvik, C., Lentz, T.E., Hedegaard, M.M., Johansen, S.D. (2009). A conformational switch in the DiGIR1 ribozyme involved in release and folding of the downstream I-DirI mRNA. *RNA*, 15(5), 958–967.

Niranjanakumari, S., Stams, T., Crary, S.M., Christianson, D.W., Fierke, C.A. (1998). Protein component of the ribozyme ribonuclease P alters substrate recognition by directly contacting precursor tRNA. *Proc. Natl. Acad. Sci. USA*, 95(26), 15212–15217.

Nissen, P., Hansen, J., Ban, N., Moore, P.B., Steitz, T.A. (2000). The structural basis of ribosome activity in peptide bond synthesis. *Science*, 289(5481), 920–930.

Nissen, P., Ippolito, J.A., Ban, N., Moore, P.B., Steitz, T.A. (2001). RNA tertiary interactions in the large ribosomal subunit: The A-minor motif. *Proc. Natl. Acad. Sci. USA*, 98(9), 4899–4903.

Ogle, J.M., Brodersen, D.E., Clemons, W.M., Tarry, M.J., Carter, A.P., Ramakrishnan, V. (2001). Recognition of cognate transfer RNA by the 30S ribosomal subunit. *Science*, 292(5518), 897–902.

Ohno, S. (1970). *Evolution by Gene Duplication*. Allen & Unwin, London.

Ohuchi, M., Murakami, H., Suga, H. (2006). In vitro evolution of flexizymes that function under the conditions in translation system. *Nucleic Acids Symp. Ser. (Oxf.)*, (50), 299–300.

O'Rear, J.L., Wang, S., Feig, A.L., Beigelman, L., Uhlenbeck, O.C., Herschlag, D. (2001). Comparison of the hammerhead cleavage reactions stimulated by monovalent and divalent cations. *RNA*, 7(4), 537–545.

Orgel, L.E. (1968). Evolution of the genetic apparatus. *J. Mol. Biol.*, 38(3), 381–393.

Orias, E., Cervantes, M.D., Hamilton, E.P. (2011). *Tetrahymena thermophila*, a unicellular eukaryote with separate germline and somatic genomes. *Res. Microbiol.*, 162(6), 578–586.

Pan, T. and Uhlenbeck, O.C. (1992). A small metalloribozyme with a two-step mechanism. *Nature*, 358, 560–563.

Pavlova, N., Kaloudas, D., Penchovsky, R. (2019). Riboswitch distribution, structure, and function in bacteria. *Gene*, 708, 38–48.

Peltier, J., Shaw, H.A., Couchman, E.C., Dawson, L.F., Yu, L., Choudhary, J.S., Kaever, V., Wren, B.W., Fairweather, N.F. (2015). Cyclic diGMP regulates production of sortase substrates of Clostridium difficile and their surface exposure through ZmpI protease-mediated cleavage. *J. Biol. Chem.*, 290(40), 24453–24469.

Penchovsky, R. (2014). Computational design of allosteric ribozymes as molecular biosensors. *Biotechnol. Adv.*, 32(5), 1015–1027.

Peracchi, A., Beigelman, L., Scott, E.C., Uhlenbeck, O.C., Herschlag, D. (1997). Involvement of a specific metal ion in the transition of the hammerhead ribozyme to its catalytic conformation. *J. Biol. Chem.*, 272(43), 26822–26826.

Perreault, J., Weinberg, Z., Roth, A., Popescu, O., Chartrand, P., Ferbeyre, G., Breaker, R.R. (2011). Identification of hammerhead ribozymes in all domains of life reveals novel structural variations. *PLoS Comput. Biol.*, 7(5), e1002031.

Perrotta, A.T., Shih, I., Been, M.D. (1999). Imidazole rescue of a cytosine mutation in a self-cleaving ribozyme. *Science*, 286(5437), 123–126.

Pham, H.L., Wong, A., Chua, N., Teo, W.S., Yew, W.S., Chang, M.W. (2017). Engineering a riboswitch-based genetic platform for the self-directed evolution of acid-tolerant phenotypes. *Nat. Commun.*, 8(1), 411.

Piccirilli, J.A. and Koldobskaya, Y. (2011). Crystal structure of an RNA polymerase ribozyme in complex with an antibody fragment. *Philos. Trans. R. Soc. Lond. B Biol. Sci.*, 366(1580), 2918–2928.

Pleij, C.W.A. (1990). Pseudoknots: A new motif in the RNA game. *TIBS*, 15(4), 143–147.

Pley, H.W., Flaherty, K.M., McKay, D.B. (1994a). Model for an RNA tertiary interaction from the structure of an intermolecular complex between a GAAA tetraloop and a n RNA helix. *Nature*, 372, 111–113.

Pley, H.W., Flaherty, K.M., McKay, D.B. (1994b). Three-dimensional structure of a hammerhead ribozyme. *Nature*, 372, 68–74.

Polacek, N. and Mankin, A.S. (2005). The ribosomal peptidyl transferase center: Structure, function, evolution, inhibition. *Crit. Rev. Biochem. Mol. Biol.*, 40(5), 285–311.

Polacek, N., Gaynor, M., Yassin, A., Mankin, A.S. (2001). Ribosomal peptidyl transferase can withstand mutations at the putative catalytic nucleotide. *Nature*, 411(6836), 498–501.

Pontius, B.W., Lott, W.B., Von Hippel, P.H. (1997). Observations on catalysis by hammerhead ribozymes are consistent with a two-divalent-metal-ion mechanism. *Proc. Natl. Acad. Sci. USA*, 94(6), 2290.

Przytula-Mally, I., Engilberge, S., Johannsen, S., Oliéric, V., Masquida, B., Sigel, R.K.O. (2022). Anticodon-like loop-mediated dimerization in the crystal structures of HdV-like CPEB3 ribozymes. Submitted.

Qu, G., Dong, X., Piazza, C.L., Chalamcharla, V.R., Lutz, S., Curcio, M.J., Belfort, M. (2014). RNA-RNA interactions and pre-mRNA mislocalization as drivers of group II intron loss from nuclear genomes. *Proc. Natl. Acad. Sci. USA*, 111(18), 6612–6617.

Qu, W.R., Sun, Q.H., Liu, Q.Q., Jin, H.J., Cui, R.J., Yang, W., Song, B., Li, B.J. (2020). Role of CPEB3 protein in learning and memory: New insights from synaptic plasticity. *Aging (Albany NY)*, 12(14), 15169–15182.

Raines, R.T. (1998). Ribonuclease A. *Chem. Rev.*, 98(3), 1045–1066.

Reeder, J. and Giegerich, R. (2004). Design, implementation and evaluation of a practical pseudoknot folding algorithm based on thermodynamics. *BMC Bioinform.*, 5, 104.

Regulski, E.E. and Breaker, R.R. (2008). In-line probing analysis of riboswitches. In *Post-Transcriptional Gene Regulation*, Wilusz, J. (ed.). Humana Press, Totowa.

Reiter, N.J., Osterman, A., Torres-Larios, A., Swinger, K.K., Pan, T., Mondragon, A. (2010). Structure of a bacterial ribonuclease P holoenzyme in complex with tRNA. *Nature*, 468(7325), 784–789.

Ren, A., Vušurović, N., Gebetsberger, J., Gao, P., Juen, M., Kreutz, C., Micura, R., Patel, D.J. (2016). Pistol ribozyme adopts a pseudoknot fold facilitating site-specific in-line cleavage. *Nat. Chem. Biol.*, 12, 702.

Richter, J.D. (2007). CPEB: A life in translation. *Trends Biochem. Sci.*, 32(6), 279–285.

Rivas, E. and Eddy, S.R. (1999). A dynamic programming algorithm for RNA structure prediction including pseudoknots. *J. Mol. Biol.*, 285(5), 2053–2068.

Rivas, E., Clements, J., Eddy, S.R. (2017). A statistical test for conserved RNA structure shows lack of evidence for structure in lncRNAs. *Nat. Methods*, 14(1), 45–48.

Robertson, D.L. and Joyce, G.F. (1990). Selection in vitro of an RNA enzyme that specifically cleaves single-stranded DNA. *Nature*, 344, 467–468.

Robertson, M.P. and Scott, W.G. (2007). The structural basis of ribozyme-catalyzed RNA assembly. *Science*, 315(5818), 1549–1553.

Robertson, H.D., Altman, S., Smith, J.D. (1972). Purification and properties of a specific *Escherichia coli* ribonuclease which cleaves a tyrosine transfer ribonucleic acid presursor. *J. Biol. Chem.*, 247(16), 5243–5251.

Roger, A.J. and Simpson, A.G. (2009). Evolution: Revisiting the root of the eukaryote tree. *Curr. Biol.*, 19(4), R165–167.

Rogers, S.O. (2019). Integrated evolution of ribosomal RNAs, introns, and intron nurseries. *Genetica*, 147(2), 103–119.

Romling, U. and Amikam, D. (2006). Cyclic di-GMP as a second messenger. *Curr. Opin. Microbiol.*, 9(2), 218–228.

Roth, A., Weinberg, Z., Chen, A.G., Kim, P.B., Ames, T.D., Breaker, R.R. (2014). A widespread self-cleaving ribozyme class is revealed by bioinformatics. *Nat. Chem. Biol.*, 10(1), 56–60.

Rozov, A., Demeshkina, N., Khusainov, I., Westhof, E., Yusupov, M., Yusupova, G. (2016a). Novel base-pairing interactions at the tRNA wobble position crucial for accurate reading of the genetic code. *Nat. Commun.*, 7, 10457.

Rozov, A., Demeshkina, N., Westhof, E., Yusupov, M., Yusupova, G. (2016b). New structural insights into translational miscoding. *Trends Biochem. Sci.*, 41(9), 798–814.

Rozov, A., Wolff, P., Grosjean, H., Yusupov, M., Yusupova, G., Westhof, E. (2018). Tautomeric G*U pairs within the molecular ribosomal grip and fidelity of decoding in bacteria. *Nucleic Acids Res.*, 46(14), 7425–7435.

Rozov, A., Khusainov, I., El Omari, K., Duman, R., Mykhaylyk, V., Yusupov, M., Westhof, E., Wagner, A., Yusupova, G. (2019). Importance of potassium ions for ribosome structure and function revealed by long-wavelength X-ray diffraction. *Nat. Commun.*, 10(1), 2519.

Rupert, P.B. and Ferré-D'Amaré, A.R. (2001). Crystal structure of a hairpin ribozyme-inhibitor complex with implications for catalysis. *Nature*, 410(6830), 780–786.

Rupert, P.B., Massey, A.P., Sigurdsson, S.T., Ferré-D'Amaré, A.R. (2002). Transition state stabilization by a catalytic RNA. *Science*, 298(5597), 1421–1424.

Ryckelynck, M., Baudrey, S., Rick, C., Marin, A., Coldren, F., Westhof, E., Griffiths, A.D. (2015). Using droplet-based microfluidics to improve the catalytic properties of RNA under multiple-turnover conditions. *RNA*, 21(3), 458–469.

Saenger, W. (1984a). *Principles of Nucleic Acid Structure*. Springer-Verlag, New York.

Saenger, W. (1984b). tRNA – A treasury of stereochemical information. In *Principles of Nucleic Acid Structure*. Springer Verlag, New York.

Saksmerprome, V., Roychowdhury-Saha, M., Jayasena, S., Khvorova, A., Burke, D.H. (2004). Artificial tertiary motifs stabilize trans-cleaving hammerhead ribozymes under conditions of submillimolar divalent ions and high temperatures. *RNA*, 10(12), 1916–1924.

Salazar, G., Paoli, L., Alberti, A., Huerta-Cepas, J., Ruscheweyh, H.-J., Cuenca, M., Field, C.M., Coelho, L.P., Cruaud, C., Engelen, S. et al. (2019). Gene expression changes and community turnover differentially shape the global ocean metatranscriptome. *Cell*, 179(5), 1068–1083, e1021.

Salehi-Ashtiani, K., Luptak, A., Litovchick, A., Szostak, J.W. (2006). A genomewide search for ribozymes reveals an HDV-like sequence in the human CPEB3 gene. *Science*, 313(5794), 1788–1792.

Sankoff, D. (1985). Simultaneous solution of the RNA folding, alignment and protosequence problems. *SIAM J. Appl. Math.*, 45(5), 810–825.

Sargueil, B. and Burke, J.M. (1997). In vitro selection of hairpin ribozymes. *Methods Mol. Biol.*, 74, 289–300.

Sargueil, B., Hampel, K.J., Lambert, D., Burke, J.M. (2003). In vitro selection of second site revertants analysis of the hairpin ribozyme active site. *J. Biol. Chem.*, 278(52), 52783–52791.

Saville, B.J. and Collins, R.A. (1990). A site-specific self-cleavage reaction performed by a novel RNA in Neurospora mitochondria. *Cell*, 61, 685–696.

Scarborough, R.J. and Gatignol, A. (2015). HIV and ribozymes. In *Gene Therapy for HIV and Chronic Infections*, Berkhout, B., Ertl, H.C.J., Weinberg, M.S. (eds). Springer, New York.

Schrodinger, L. (2010). *The PyMOL molecular graphics system*, Version 1.3r1. Manuscript.

Schultz, L.W., Quirk, D.J., Raines, R.T. (1998). His...Asp catalytic dyad of ribonuclease A: Structure and function of the wild-type, D121N, and D121A enzymes. *Biochemistry*, 37(25), 8886–8898.

Scott, W.G., Finch, J.T., Grenfell, R., Fogg, J., Smith, T., Gait, M.J., Klug, A. (1995a). Rapid crystallization of chemically synthesized Hammerhead RNAs using a double screening procedure. *J. Mol. Biol.*, 250, 327–332.

Scott, W.G., Finch, J.T., Klug, A. (1995b). The crystal structure of an all-RNA hammerhead ribozyme: A proposed mechanism for RNA catalytic cleavage. *Cell*, 81, 991–1002.

Scott, W.G., Murray, J.B., Arnold, J.R.P., Stoddard, B.L., Klug, A. (1996). Capturing the structure of a catalytic RNA intermediate: The hammerhead ribozyme. *Science*, 274, 2065–2069.

Scott, W.G., Martick, M., Chi, Y.I. (2009). Structure and function of regulatory RNA elements: Ribozymes that regulate gene expression. *Biochim. Biophys. Acta*, 1789(9/10), 634–641.

Serganov, A., Keiper, S., Malinina, L., Tereshko, V., Skripkin, E., Hobartner, C., Polonskaia, A., Phan, A.T., Wombacher, R., Micura, R. et al. (2005). Structural basis for Diels–Alder ribozyme-catalyzed carbon–carbon bond formation. *Nat. Struct. Mol. Biol.*, 12(3), 218–224.

Shechner, D.M. and Bartel, D.P. (2011). The structural basis of RNA-catalyzed RNA polymerization. *Nat. Struct. Mol. Biol.*, 18(9), 1036–1042.

Shechner, D.M., Grant, R.A., Bagby, S.C., Koldobskaya, Y., Piccirilli, J.A., Bartel, D.P. (2009). Crystal structure of the catalytic core of an RNA-polymerase ribozyme. *Science*, 326(5957), 1271–1275.

Singh, V., Fedeles, B.I., Essigmann, J.M. (2015). Role of tautomerism in RNA biochemistry. *RNA*, 21(1), 1–13.

Smith, M.A., Gesell, T., Stadler, P.F., Mattick, J.S. (2013). Widespread purifying selection on RNA structure in mammals. *Nucleic Acids Res.*, 41(17), 8220–8236.

Soukup, G.A. (2006). Allosteric ribozymes as molecular switches and sensors. In *Nucleic Acid Switches and Sensors*, Silverman, S.K. (ed.). Springer US, Boston.

Sripathi, K.N., Banas, P., Reblova, K., Sponer, J., Otyepka, M., Walter, N.G. (2015). Wobble pairs of the HDV ribozyme play specific roles in stabilization of active site dynamics. *Phys. Chem. Phys.*, 17(8), 5887–5900.

Stams, T., Niranjanakumari, S., Fierke, C.A., Christianson, D.W. (1998). Ribonuclease P protein structure: Evolutionary origins in the translational apparatus. *Science*, 280(5364), 752–755.

Stark, B.C., Kole, R., Bowman, E.J., Altman, S. (1978). Ribonuclease P: An enzyme with an essential RNA component. *Proc. Natl. Acad. Sci. USA*, 75(8), 3717–3721.

Statello, L., Guo, C.J., Chen, L.L., Huarte, M. (2021). Gene regulation by long non-coding RNAs and its biological functions. *Nat. Rev. Mol. Cell. Biol.*, 22(2), 96–118.

Steitz, T.A. and Steitz, J.A. (1993). A general two-metal-ion mechanism for catalytic RNA. *Proc. Natl. Acad. Sci. USA*, 90(14), 6498–6502.

Stephan, J.S., Fioriti, L., Lamba, N., Colnaghi, L., Karl, K., Derkatch, I.L., Kandel, E.R. (2015). The CPEB3 protein is a functional prion that interacts with the actin cytoskeleton. *Cell Rep.*, 11(11), 1772–1785.

Stombaugh, J., Zirbel, C.L., Westhof, E., Leontis, N.B. (2009). Frequency and isostericity of RNA base pairs. *Nucleic Acids Res.*, 37(7), 2294–2312.

Suga, H., Lohse, P.A., Szostak, J.W. (1998). Structural and kinetic characterization of an acyl transferase ribozyme. *J. Am. Chem. Soc.*, 120(6), 1151–1156.

Sundaralingam, M. (1973). The concept of a conformationally "rigid" nucleotide and its significance in polynucleotide conformational analysis. In *Conformation of Biological Molecules and Polymers*, Pullmann, V.E.D.B.B. (ed.). The Israel Academy of Sciences and Humanities, Jerusalem.

Sureau, C. and Negro, F. (2016). The hepatitis delta virus: Replication and pathogenesis. *J. Hepatol.*, 64(1 Suppl), S102–S116.

Suslov, N.B., Dasgupta, S., Huang, H., Fuller, J.R., Lilley, D.M., Rice, P.A., Piccirilli, J.A. (2015). Crystal structure of the Varkud satellite ribozyme. *Nat. Chem. Biol.*, 11(11), 840–846.

Symons, R.H. (1989). Self-cleavage of RNA in the replication of small pathogens of plants and animals. *Trends Biochem. Sci.*, 14(11), 445–450.

Szewczak, L.B., Gabrielsen, J.S., Degregorio, S.J., Strobel, S.A., Steitz, J.A. (2005). Molecular basis for RNA kink-turn recognition by the h15.5K small RNP protein. *RNA*, 11(9), 1407–1419.

Taft, R.J., Pheasant, M., Mattick, J.S. (2007). The relationship between non-protein-coding DNA and eukaryotic complexity. *Bioessays*, 29(3), 288–299.

Taira, K., Uebayasi, M., Furukawa, K. (1989). Cyclic oxyphosphoranes as model intermediates during splicing and cleavage of RNA: Ab initio molecular orbital calculations on the conformational analysis. *Nucleic Acids Res.*, 17(10), 3699–3708.

Takahashi, C., Sheng, Z., Horan, T.P., Kitayama, H., Maki, M., Hitomi, K., Kitaura, Y., Takai, S., Sasahara, R.M., Horimoto, A. et al. (1998). Regulation of matrix metalloproteinase-9 and inhibition of tumor invasion by the membrane-anchored glycoprotein RECK. *Proc. Natl. Acad. Sci. USA*, 95(22), 13221–13226.

Talini, G., Gallori, E., Maurel, M.C. (2009). Natural and unnatural ribozymes: Back to the primordial RNA world. *Res. Microbiol.*, 160(7), 457–465.

Tang, Y., Nielsen, H., Masquida, B., Gardner, P.P., Johansen, S.D. (2014). Molecular characterization of a new member of the lariat capping twin-ribozyme introns. *Mob. DNA*, 5, 25.

Thompson, J.E., Kutateladze, T.G., Schuster, M.C., Venegas, F.D., Messmore, J.M., Raines, R.T. (1995). Limits to catalysis by ribonuclease A. *Bioorg. Chem.*, 23(4), 471–481.

Tjhung, K.F., Shokhirev, M.N., Horning, D.P., Joyce, G.F. (2020). An RNA polymerase ribozyme that synthesizes its own ancestor. *Proc. Natl. Acad. Sci. USA*, 117(6), 2906–2913.

Tocchini-Valentini, G.D., Fruscoloni, P., Tocchini-Valentini, G.P. (2011). Evolution of introns in the archaeal world. *Proc. Natl. Acad. Sci. USA*, 108(12), 4782–4787.

Toor, N., Keating, K.S., Taylor, S.D., Pyle, A.M. (2008a). Crystal structure of a self-spliced group II intron. *Science*, 320(5872), 77–82.

Toor, N., Rajashankar, K., Keating, K.S., Pyle, A.M. (2008b). Structural basis for exon recognition by a group II intron. *Nat. Struct. Mol. Biol.*, 15(11), 1221–1222.

Torres-Larios, A., Swinger, K.K., Krasilnikov, A.S., Pan, T., Mondragon, A. (2005). Crystal structure of the RNA component of bacterial ribonuclease P. *Nature*, 437(7058), 584–587.

Torres-Larios, A., Swinger, K.K., Pan, T., Mondragon, A. (2006). Structure of ribonuclease P – A universal ribozyme. *Curr. Opin. Struct. Biol.*, 16(3), 327–335.

Tsiamantas, C., Otero-Ramirez, M.E., Suga, H. (2019). Discovery of functional macrocyclic peptides by means of the RaPID system. In *Cyclic Peptide Design*, Goetz, G. (ed). Springer, New York.

Tucker, B.J. and Breaker, R.R. (2005). Riboswitches as versatile gene control elements. *Curr. Opin. Struct. Biol.*, 15(3), 342.

Tuerk, C. and Gold, L. (1990). Systematic evolution of ligands by exponential enrichment: RNA ligands to bacteriophage T4 DNA polymerase. *Science*, 249, 505–510.

Turk, E.M., Das, V., Seibert, R.D., Andrulis, E.D. (2013). The mitochondrial RNA landscape of *Saccharomyces cerevisiae*. *PLoS One*, 8(10), e78105.

Uchimaru, T., Storer, J.W., Tanabe, K., Uebayasi, M., Nishikawa, S., Taira, K. (1992). RNA hydrolysis via an oxyphosphorane intermediate. *Biochem. Biophys. Res. Commun.*, 187(3), 1523–1528.

Uchimaru, T., Tanabe, K., Shimayama, T., Uebayasi, M., Taira, K. (1993a). Theoretical analyses on the role of metal cations in RNA cleavage processes. *Nucleic Acids Symp. Ser.*, 29, 179–180.

Uchimaru, T., Uebayasi, M., Tanabe, K., Taira, K. (1993b). Theoretical analyses on the role of Mg2+ ions in ribozyme reactions. *FASEB J.*, 7(1), 137–142.

Uchimaru, T., Uebayasi, M., Hirose, T., Tsuzuki, S., Yliniemela, A., Tanabe, K., Taira, K. (1996). Electrostatic interactions that determine the rate of pseudorotation processes in oxyphosphorane intermediates: Implications with respect to the roles of metal ions in the enzymatic cleavage of RNA. *J. Org. Chem.*, 61(5), 1599–1608.

Uhlenbeck, O.C. (1987). A small catalytic oligoribonucleotide. *Nature*, 328(6131), 596–600.

Vader, A., Nielsen, H., Johansen, S. (1999). In vivo expression of the nucleolar group I intron-encoded I-dirI homing endonuclease involves the removal of a spliceosomal intron. *EMBO J.*, 18(4), 1003–1013.

Vader, A., Johansen, S., Nielsen, H. (2002). The group I-like ribozyme DiGIR1 mediates alternative processing of pre-rRNA transcripts in *Didymium iridis*. *Eur. J. Biochem.*, 269(23), 5804–5812.

Vidovic, I., Nottrott, S., Hartmuth, K., Luhrmann, R., Ficner, R. (2000). Crystal structure of the spliceosomal 15.5kD protein bound to a U4 snRNA fragment. *Mol. Cell*, 6(6), 1331–1342.

Vogler, C., Spalek, K., Aerni, A., Demougin, P., Muller, A., Huynh, K.D., Papassotiropoulos, A., De Quervain, D.J. (2009). CPEB3 is associated with human episodic memory. *Front Behav. Neurosci.*, 3(4).

Waldsich, C., Masquida, B., Westhof, E., Schroeder, R. (2002). Monitoring intermediate folding states of the td group I intron in vivo. *EMBO J.*, 21(19), 5281–5291.

Wang, J.Y., Pausch, P., Doudna, J.A. (2022). Structural biology of CRISPR – Cas immunity and genome editing enzymes. *Nat. Rev. Microbiol.*, 20, 641–656.

Ward, W.L. and Derose, V.J. (2012). Ground-state coordination of a catalytic metal to the scissile phosphate of a tertiary-stabilized Hammerhead ribozyme. *RNA*, 18(1), 16–23.

Waterhouse, A.M., Procter, J.B., Martin, D.M., Clamp, M., Barton, G.J. (2009). Jalview Version 2 – A multiple sequence alignment editor and analysis workbench. *Bioinformatics*, 25(9), 1189–1191.

Watson, J.D. and Crick, F.H. (1953). Molecular structure of nucleic acids; a structure for deoxyribose nucleic acid. *Nature*, 171(4356), 737–738.

Webb, C.H. and Luptak, A. (2011). HDV-like self-cleaving ribozymes. *RNA Biol.*, 8(5), 719–727.

Webb, C.H., Riccitelli, N.J., Ruminski, D.J., Luptak, A. (2009). Widespread occurrence of self-cleaving ribozymes. *Science*, 326(5955), 953.

Wedekind, J.E. and McKay, D.B. (2003). Crystal structure of the leadzyme at 1.8 A resolution: Metal ion binding and the implications for catalytic mechanism and allo site ion regulation. *Biochemistry*, 42(32), 9554–9563.

Weinberg, C.E. (2021). Biological roles of self-cleaving ribozymes. In *Ribozymes*, Sabine Müller, B.M. and Winkler, W. (eds). Wiley-VCH, Weinheim.

Weinberg, Z. and Ruzzo, W.L. (2004). Exploiting conserved structure for faster annotation of non-coding RNAs without loss of accuracy. *Bioinformatics*, 20(Suppl 1), i334–341.

Weinberg, Z. and Ruzzo, W.L. (2006). Sequence-based heuristics for faster annotation of non-coding RNA families. *Bioinformatics*, 22(1), 35–39.

Weinberg, Z., Barrick, J.E., Yao, Z., Roth, A., Kim, J.N., Gore, J., Wang, J.X., Lee, E.R., Block, K.F., Sudarsan, N. et al. (2007). Identification of 22 candidate structured RNAs in bacteria using the CMfinder comparative genomics pipeline. *Nucl. Acids Res.*, 35(14), 4809–4819.

Weinberg, Z., Wang, J.X., Bogue, J., Yang, J., Corbino, K., Moy, R.H., Breaker, R.R. (2010). Comparative genomics reveals 104 candidate structured RNAs from bacteria, archaea, and their metagenomes. *Genome Biol.*, 11(3), R31.

Weinberg, Z., Kim, P.B., Chen, T.H., Li, S., Harris, K.A., Lunse, C.E., Breaker, R.R. (2015). New classes of self-cleaving ribozymes revealed by comparative genomics analysis. *Nat. Chem. Biol.*, 11(8), 606–610.

Weinberg, C.E., Olzog, V.J., Eckert, I., Weinberg, Z. (2021). Identification of over 200-fold more hairpin ribozymes than previously known in diverse circular RNAs. *Nucleic Acids Res.*, 49(11), 6375–6388.

Westhof, E. and Jaeger, L. (1992). RNA pseudoknots: Structural and functional aspects. *Curr. Opin. Struct. Biol.*, 2, 327–333.

Wikmark, O.-G., Einvik, C., De Jonckheere, J., Johansen, S. (2006). Short-term sequence evolution and vertical inheritance of the Naegleria twin-ribozyme group I intron. *BMC Evol. Biol.*, 6(1), 39–50.

Wilkinson, M.E., Charenton, C., Nagai, K. (2020). RNA splicing by the spliceosome. *Annu. Rev. Biochem.*, 89, 359–388.

Wilson, T.J., McLeod, A.C., Lilley, D.M. (2007). A guanine nucleobase important for catalysis by the VS ribozyme. *EMBO J.*, 26(10), 2489–2500.

Wilson, T.J., Liu, Y., Domnick, C., Kath-Schorr, S., Lilley, D.M. (2016). The novel chemical mechanism of the twister ribozyme. *J. Am. Chem. Soc.*, 138(19), 6151–6162.

Wilson, T.J., Liu, Y., Li, N.S., Dai, Q., Piccirilli, J.A., Lilley, D.M. (2019). Comparison of the structures and mechanisms of the pistol and hammerhead ribozymes. *J. Am. Chem. Soc.*, 141(19), 7865–7875.

Wimberly, B.T., Brodersen, D.E., Clemons, W.M., Morgan-Warren, R.J., Carter, A.P., Vonrhein, C., Hartsch, T., Ramakrishnan, V. (2000). Structure of the 30S ribosomal subunit. *Nature*, 407(6802), 327–339.

Winkler, W.C., Nahvi, A., Sudarsan, N., Barrick, J.E., Breaker, R.R. (2003). An mRNA structure that controls gene expression by binding S-adenosylmethionine. *Nat. Struct. Biol.*, 10(9), 701–707.

Winkler, W.C., Nahvi, A., Roth, A., Collins, J.A., Breaker, R.R. (2004). Control of gene expression by a natural metabolite-responsive ribozyme. *Nature*, 428(6980), 281–286.

Woese, C.R. (1967). *The Genetic Code: The Molecular Basis for Genetic Expression*. Harper & Row, New York.

Woodson, S.A. (2010). Compact intermediates in RNA folding. *Annu. Rev. Biophys.*, 39, 61–77.

Xayaphoummine, A., Bucher, T., Thalmann, F., Isambert, H. (2003). Prediction and statistics of pseudoknots in RNA structures using exactly clustered stochastic simulations. *Proc. Natl. Acad. Sci. USA*, 100(26), 15310–15315.

Xiao, H., Murakami, H., Suga, H., Ferré-D'Amaré, A.R. (2008). Structural basis of specific tRNA aminoacylation by a small in vitro selected ribozyme. *Nature*, 454(7202), 358–361.

Yan, C., Hang, J., Wan, R., Huang, M., Wong, C.C., Shi, Y. (2015). Structure of a yeast spliceosome at 3.6-angstrom resolution. *Science*, 349(6253), 1182–1191.

Yusupov, M.M., Yusupova, G.Z., Baucom, A., Lieberman, K., Earnest, T.N., Cate, J.H., Noller, H.F. (2001). Crystal structure of the ribosome at 5.5. A resolution. *Science*, 292(5518), 883–896.

Yusupova, G.Z., Yusupov, M.M., Cate, J.H., Noller, H.F. (2001). The path of messenger RNA through the ribosome. *Cell*, 106(2), 233–241.

Zaug, A.J. and Cech, T.R. (1986a). The intervening sequence RNA of *Tetrahymena* is an enzyme. *Science*, 231(4737), 470–475.

Zaug, A.J. and Cech, T.R. (1986b). The *Tetrahymena* intervening sequence ribonucleic acid enzyme is a phosphotransferase and an acid phosphatase. *Biochemistry*, 25(16), 4478–4482.

Zaug, A.J., Been, M.D., Cech, T.R. (1986). The *Tetrahymena* ribozyme acts like an RNA restriction endonuclease. *Nature*, 324(6096), 429–433.

Zhang, X., Yan, C., Hang, J., Finci, L.I., Lei, J., Shi, Y. (2017). An atomic structure of the human spliceosome. *Cell*, 169(5), 918–929, e914.

Zheng, H., Shabalin, I.G., Handing, K.B., Bujnicki, J.M., Minor, W. (2015). Magnesium-binding architectures in RNA crystal structures: Validation, binding preferences, classification and motif detection. *Nucleic Acids Res.*, 43(7), 3789–3801.

Zheng, L., Falschlunger, C., Huang, K., Mairhofer, E., Yuan, S., Wang, J., Patel, D.J., Micura, R., Ren, A. (2019). Hatchet ribozyme structure and implications for cleavage mechanism. *Proc. Natl. Acad. Sci. USA*, 116(22), 10783–10791.

Zhou, L., Hang, J., Zhou, Y., Wan, R., Lu, G., Yin, P., Yan, C., Shi, Y. (2014). Crystal structures of the Lsm complex bound to the 3' end sequence of U6 small nuclear RNA. *Nature*, 506(7486), 116–120.

Zorzan, M., Giordan, E., Redaelli, M., Caretta, A., Mucignat-Caretta, C. (2015). Molecular targets in glioblastoma. *Future Oncol.*, 11(9), 1407–1420.

Zuker, M. (1989). Computer prediction of RNA structure. *Methods Enzymol.*, 180, 262–288.

Zuker, M. (2003). Mfold web server for nucleic acid folding and hybridization prediction. *Nucleic Acids Res.*, 31(13), 3406–3415.

Zwanzig, R., Szabo, A., Bagchi, B. (1992). Levinthal's paradox. *Proc. Natl. Acad. Sci. USA*, 89, 20–22.

Index

A, B

A-minor, 20–24
AA platform, 19
adenine, 3, 19, 21, 52, 95
adenosine, 20, 32, 39, 46, 55, 56, 59
allosteric, 77
Alu, 124, 130
anticodon, 22–24, 32, 45, 47, 51, 53, 54
antiparallel, 2–4, 6, 8, 21
biofilm, 77
branch point, 55

C, D

catalysis, 1, 5, 29, 37, 38, 42, 46, 49, 52, 55, 56, 69, 73, 79, 85, 88, 90, 91, 95, 99, 102–106, 123, 126, 129
central dogma, 29
codon, 22, 23, 32, 49, 51, 54, 78
cofactor, 46, 77, 89–91
concatamer, 118
coronavirus, 29
CPEB3, 63, 76, 125–129, 131
cytosine, 3, 127
decoding, 22, 23, 29, 47, 51
didymium, 91, 93, 94
dihedral, 6, 7
double-stranded, 2–5, 8, 59, 120

E, G

electrophilic, 34, 70, 82, 95
endonuclease, 62, 75, 93–95, 97, 116, 117, 121, 124
enzyme, 30, 37, 38, 42, 43, 45, 59, 60, 69, 73, 90, 94, 99, 103, 109, 121
epigenetic, 35, 59, 125
exon, 31, 41, 55, 56, 128
genome, 29, 35, 62, 63, 71, 111, 113, 114, 117, 120, 124, 126, 130
glmS, 69, 77, 86, 89–91, 128
group, 2, 3, 5, 7, 14, 16, 19–21, 31, 34, 38–41, 43, 46, 49, 52–54, 56, 59, 69, 70, 74–78, 82, 84–86, 88, 90, 91, 93–95, 97, 99–104, 106–109, 111, 112, 115, 117, 123, 127
guanine, 16, 51, 59, 66, 107, 127
guanosine, 31, 39, 46, 51, 54, 59, 91

H, I

hairpin, 16, 54, 59, 60, 66, 78, 84, 86–88, 95, 126, 128
hammerhead, 18, 20, 26, 59, 60, 63, 66, 69, 78, 81–86, 88, 99, 101, 103–105, 107, 108, 114, 117–121, 123, 126, 129

hatchet, 69, 78, 108, 121, 123
helix, 3, 5, 6, 8, 11–13, 16, 18, 20, 22, 24, 26, 41, 51, 52, 66, 69, 97, 114, 126, 128
hepatitis, 29, 52, 59, 63, 66, 107
herpes, 29
Hoogsteen, 4, 5, 11, 16, 19, 54
Hovlinc, 129, 130
imino, 50, 52, 107
immunodeficiency, 29
initiator, 77
intron, 16, 19–21, 30, 31, 37, 39, 41, 42, 46, 55, 56, 77, 91–94, 97, 115, 117, 123, 124, 126, 127

J, K, L

junction, 8, 62, 66, 82, 88, 94, 95, 97, 128
k-turn, 16–18, 24
lariat, 39, 55, 56, 91–93, 95, 97, 116, 117, 128
lasso, 39, 42, 75
loop E, 14–16, 24
LTR (Long Term Repeat), 120

M, N

messenger, 22, 23, 29, 34, 47, 51, 53–56, 63, 75, 77, 90, 95, 99, 115–117, 120, 123, 124, 127, 128
metabolite, 69
motif, 10, 12, 14–18, 20, 22–24, 118, 126
mRNA, 18, 22, 23, 32, 47, 49, 120
Naegleria, 91, 97
NGS (Next Generation Sequencing), 25, 54, 62, 69
nucleobase, 2, 6, 52, 54, 127
nucleolytic, 52, 56, 63, 67–69, 73, 75, 78, 103, 105, 107

nucleotide, 1–3, 5, 6, 14, 16, 17, 20, 21, 24, 31, 32, 34, 38, 39, 41, 50, 51, 53–55, 59, 73, 86–88, 91, 93, 95, 104, 108, 114, 116, 117, 127

O, P

oxyphosphorane, 34, 59, 82, 84, 86, 88, 100, 107
PCR (Polymerase Chain Reaction), 57, 62, 72, 73, 78
phosphate, 2–6, 8, 11, 31, 34, 41, 42, 57, 59, 66, 69, 70, 73, 77, 85, 89–91, 93, 95, 99–101, 104–109, 118
pistol, 69, 78, 108
polyadenylation, 63, 126
polyanion, 2, 3
Pospiviroidae, 117
proofreading, 22
pseudoknot, 12–14, 97, 116, 117, 126
purine, 3, 5
pyrimidine, 3, 6

R, S

RBS (ribosomal binding site), 77
recombination, 34, 88, 93, 94, 111
RFam, 25, 26, 75, 76
ribosome, 12, 22, 24, 32, 78, 95, 103, 115
riboswitch, 14, 63, 77, 86, 89, 90, 109
right-handed, 2, 3, 5
RNA world, 34, 56, 71, 78, 113, 123
RNAcentral, 26
RNase, 18, 31, 32, 34, 37, 42–44, 46, 52, 56, 59, 60, 74–77, 90, 99, 101–103, 111, 113, 115, 118, 128
rolling circle, 57, 59, 61, 118
rRNA (ribosomal RNA), 16, 113
Saccharomyces cerevisiae, 39, 115
satellite, 57, 59, 60, 66

sequencing, 25, 62, 78, 130
single-stranded, 8, 20, 22, 117
snRNA (small nuclear RNA), 18, 55
spliceosome, 12, 31, 32, 55, 56, 66, 75, 115
splicing, 18, 29–31, 38, 41, 46, 55, 56, 70, 75–77, 91, 95, 97, 111, 115, 117, 120, 123, 128
stereochemistry, 1, 3
sugar, 5, 11, 12, 16, 19
SUMO (Small Ubiquitin-like Modifier), 127

T, U, V

tetra-loop, 18, 20, 22, 45
Tetrahymena thermophila, 19–21, 37, 38
transcriptase, 121, 124
transesterification, 30, 31, 37, 39, 41, 56, 57, 60, 63, 69, 70, 76, 95, 99, 100
translation, 16, 23, 35, 37, 47, 51, 54, 70, 74, 77, 95, 109, 114, 120
transpeptidylation, 23, 47, 49
transposon, 120
trigonal bipyramid, 34, 59, 82, 84, 100, 101
tRNA (transfer RNA), 22, 23, 32, 37, 42, 43, 45, 47, 49, 51–54, 60, 74, 113, 115, 120
twister, 63, 69, 78, 107
  sister, 69, 107
uracil, 16, 19
uridine, 51, 54, 102, 127
vanadate, 84, 88
Varkud, 60, 66
viroid, 57, 59, 60, 69, 74, 77, 117
virulence, 77
virus, 29, 35, 52, 56, 57, 59, 63, 66, 76, 84, 107, 125

Other titles from

in

Biology and Biomedical Engineering

2023

VERNA Emeline
Asymptomatic Osseous Variations of the Postcranial Human Skeleton
(Comparative Anatomy and Posture of Animal and Human Set – Volume 5)

WALCH Jean-Paul, BLAISE Solange
Phyllotaxis Models: A Tool for Evolutionary Biologists

2022

DAMBRICOURT MALASSÉ Anne
Embryogeny and Phylogeny of the Human Posture 2: A New Glance at the Future of our Species
(Comparative Anatomy and Posture of Animal and Human Set – Volume 4)

FOSSÉ Philippe
Structures and Functions of Retroviral RNAs: The Multiple Facets of the Retroviral Genome
(Nucleic Acids Set – Volume 1)

2021

DAMBRICOURT MALASSÉ Anne
Embryogeny and Phylogeny of the Human Posture 1: A New Glance at the Future of our Species
(Comparative Anatomy and Posture of Animal and Human Set – Volume 3)

GRANDCOLAS Philippe, MAUREL Marie-Christine
Systematics and the Exploration of Life

HADJOUIS Djillali
The Skull of Quadruped and Bipedal Vertebrates: Variations, Abnormalities and Joint Pathologies
(Comparative Anatomy and Posture of Animal and Human Set – Volume 2)

HULLÉ Maurice, VERNON Philippe
The Terrestrial Macroinvertebrates of the Sub-Antarctic Îles Kerguelen and Île de la Possession

2020

CAZEAU Cyrille
Foot Surgery Viewed Through the Prism of Comparative Anatomy: From Normal to Useful
(Comparative Anatomy and Posture of Animal and Human Set – Volume 1)

DUJON Bernard, PELLETIER Georges
Trajectories of Genetics

2019

BRAND Gérard
Discovering Odors

BUIS Roger
Biology and Mathematics: History and Challenges

2016

CLERC Maureen, BOUGRAIN Laurent, LOTTE Fabien
Brain-Computer Interfaces 1: Foundations and Methods
Brain-Computer Interfaces 2: Technology and Applications

FURGER Christophe
Live Cell Assays: From Research to Health and Regulatory Applications

2015

CLARYSSE Patrick, FRIBOULET Denis
Multi-modality Cardiac Imaging: Processing and Analysis

2014

CHÈZE Laurence
Kinematic Analysis of Human Movement

DAO Tien Tuan, HO BA THO Marie-Christine
Biomechanics of the Musculoskeletal: Modeling of Data Uncertainty and Knowledge

FANET Hervé
Medical Imaging Based on Magnetic Fields and Ultrasounds

FARINAS DEL CERRO Luis, INOUE Katsumi
Logical Modeling of Biological Systems

MIGONNEY Véronique
Biomaterials

TEBBANI Sihem, LOPES Filipa, FILALI Rayen, DUMUR Didier, PAREAU Dominique
CO_2 Biofixation by Microalgae: Modeling, Estimation and Control

Printed and bound by CPI Group (UK) Ltd, Croydon, CR0 4YY
24/03/2024

14474874-0001